HISTORY OF DIPLOMACY AND TECHNOLOGY

From smoke signals to artificial intelligence

Jovan Kurbalija

Impressum

HISTORY OF DIPLOMACY AND TECHNOLOGY
From smoke signals to artificial intelligence

Publisher:
DiploFoundation (December 2023)
www.diplomacy.edu
diplo@diplomacy.edu

Author:
Jovan Kurbalija

Editing:
Mary Murphy, Paul Blamire

Design and layout:
Viktor Mijatović, Aleksandar Nedeljkov

All images in the book have been sourced responsibly. They were are either derived from the public domain or obtained with the appropriate Creative Commons licenses. Every effort has been made to respect copyright and intellectual property rights.

Except where otherwise noted, this work is licensed under
http://creativecommons.org/licences/by-nc-nd/3.0/

ISBN: 979-8-9870052-7-9

Contents

Preface	5
Introduction	9
1. Prehistory: The early origins of diplomacy and 'technologies'	13
2. Ancient Era: Cradle of humanity and diplomacy	21
3. Ancient Greece: Politics, negotiations, and diplomacy	29
4. Byzantine diplomacy: The elixir of strategic longevity	41
5. Renaissance diplomacy: Compromise as modus vivendi	47
6. Telegraph diplomacy: The 'end' of distance	61
7. Telephone diplomacy: Dialling the 'red line'	75
8. Public diplomacy: Going live on TV and radio	85
9. Digital diplomacy: Internet, AI, and social media	95
Conclusion	103
Bibliography	107
About the Author	117

Preface

The idea to write this book was triggered by my work on the impact of digital technology on diplomacy: from the advent of computers in the 1980s through the internet revolution of the 1990s to the rise of social media in the 2000s, and beyond.

Every 'latest' digital technology came with a promise of revolutionising diplomacy. Some changes happened, but the core of diplomacy remained the same: the peaceful resolution of disputes through negotiation and mediation.

I was wondering if it was the same with other technologies in history. The more I dove into history, the more I discovered this interplay between the changes that technology brought and the continuity of the core diplomatic functions. Every latest technology since smoke signals has reinforced the essence of diplomacy: representation and negotiation.

At each historical juncture, I was particularly intrigued to explore diplomatic practices in civilisations and cultures often overlooked in traditional history literature.

Now, as we stand at the threshold of the artificial intelligence (AI) era, understanding the historical backdrop that has led us here is more important than ever. In an era dominated by AI, diplomacy must develop international norms that reflect a mosaic of ethical, religious, and cultural traditions.

This book evolved from a chapter of my PhD thesis in 2005. Certain sections draw from previously published online texts of mine based on a series of discussions on the history of technology and diplomacy held in 2014 and 2021.

The insights and perspectives of the students and lecturers at Diplo have been invaluable, broadening my understanding of diplomacy and technology, particularly in the contexts of Asian, African, and indigenous traditions. A special word of thanks to Mina Mudrić, whose timely encouragement and support have been pivotal.

As you embark on this historical exploration, I encourage you to reflect on the past, engage with the present, and ponder the future trajectory of diplomacy and technology.

To my Saša

Introduction

Technology and diplomacy: A dynamic dance

Technology and diplomacy have always been intertwined. From the invention of the wheel to the development of the internet, technology has shaped the way we communicate, travel, and interact with each other. And it has had a profound impact on diplomacy.

Change and continuity have played pivotal roles in the evolution of technology and diplomacy. The fast-paced evolution of technology stands in stark contrast to the enduring essence of diplomacy – the peaceful resolution of conflicts.

In this book, we explore the fascinating interplay between technological change and diplomatic continuity. We examine the myriad ways in which technology has shaped and continues to shape diplomacy, focusing on three key areas:

- The geopolitical environment in which diplomacy operates: Technology has impacted power distribution, geopolitics, and the central relevance of cities, countries, and continents. For example, the invention of the printing press helped spread new ideas and information, which led to the Protestant Reformation, the French Revolution, and fundamental shifts in geopolitics and diplomacy. In our time, AI and digital development are triggering new redistributions of power and fundamental shifts in international relations.

- The topics that diplomats address: Technology has introduced new issues onto the diplomatic agenda. For example, the invention of the telegraph led to the need for international agreements on telecommunications regulation. And the development of AI is raising new questions about the future of warfare and international security.

- The tools that diplomats use: Technology has provided diplomats with new tools to communicate, negotiate, and build relationships. For example, the invention of the telephone allowed diplomats to hold long-distance conversations. And the development of social media has given diplomats a new way to reach out to the public and build support for their goals.

These three pillars – environment, topics, and tools – provide the backbone of our story of the evolution of diplomacy and technology. They help us to understand how technology has shaped diplomacy over time, and how it will continue to do so in the future.

We travel through history and watch as technological advancement drives both subtle changes and dramatic revolutions in diplomatic practice.

We begin with prehistoric society, where proto-diplomacy, alongside the first technologies, trade, and art, began to take shape.

We then move to the Ancient civilisations that invented writing, which has been used till today for diplomatic communication, peace agreements, and records of diplomatic interactions.

Next, we look at the diplomacy of Ancient Greece, focusing on several technological developments that played a role in diplomacy during this period, such as cryptography and the hydraulic telegraph, among others.

From there, we focus on the Byzantine Empire, one of the most enduring and sophisticated diplomatic systems that spanned over 1,000 years in the Mediterranean Basin and Anatolian Peninsula.

The next major stop on our journey is Renaissance Italy, the birthplace of diplomacy we know today with the establishment of the first permanent diplomatic missions and diplomatic archives. In the same period in the north, the printing press was established, and the Protestant Reformation set the stage for the emergence of modern diplomacy.

With the stage set, we look at the era of the telegraph, a technological breakthrough that transformed diplomacy. We see how the telegraph changed both the role of and the tools available to diplomats in the modern era.

Following the telegraph era came the arrival of the telephone, which allowed more immediate and direct communication between heads of state and diplomats.

Then came the development and proliferation of radio and television technology, and their significant diplomatic impact in the 20th century.

Finally, we discuss the transformation of diplomatic practice that has come with the spread of both the internet and social media and their role in public diplomacy, the conduct of negotiations, and other functions of diplomacy.

This historical journey is not a straight one. We make many detours to better understand the intellectual and cultural context of the evolution of technology and diplomacy. For instance, the German philosopher Karl Jasper described the Axial Age as being crucial for fostering the main philosophical and religious ideas that continue to have relevance today. This period influenced major belief systems such as Christianity, Greek philosophy, Hinduism, and Buddhism.[1]

As we explore the evolution of technology and diplomacy, we continually circle back to one enduring theme: continuity. Amid the relentless march of technological progress and the resultant changes in diplomatic practice, the fundamental ethos of diplomacy endures: the pursuit of peace through dialogue and negotiation. This book presents a journey of change, continuity, adaptation, and resilience that reflects the very essence of diplomacy.

It also serves as a reminder that no matter how much our tools and environments evolve, the core objective of diplomacy – to promote understanding, resolve conflicts, and foster peace – remains steadfast and immutable. As we turn the page to the next chapter, we delve into the fascinating annals of this ever-evolving interplay.

---

[1] Jaspers, K. (1953). *The origin and goal of history*. Yale University Press.

1. Prehistory: The early origins of diplomacy and 'technologies'

When did diplomacy start?

To find out how diplomacy began, we need to go back to prehistoric times, meaning the era before the invention of systems of writing, and look at the developments that nurtured proto-diplomacy. The behavioural sciences show that cooperation and peaceful conflict resolution are crucial for the survival and prosperity of a group. Humans most likely started solving conflicts peacefully when they developed certain cognitive abilities, self-awareness, and collective intentionality for the group they belonged to.[2] Once people developed adequate cognitive abilities, they began living in organised groups, using new technologies (making and controlling fire, and stone tools), and trading with each other.

Primatologist Prof. Frans de Waal argues that the process of bringing parties together to negotiate, i.e. diplomacy, predates the human species. His research shows that primates negotiate and mediate based on their feelings of empathy, fairness, and group interests. He further argues that the human element of diplomacy lies in the use of language.[3] These prehuman origins of diplomacy should inspire us to re-examine even the most common postulates, such as Thomas Hobbes's theory on human nature, which states that humans are predestined for conflict due to their biological drive to propagate genes.[4]

Which developments influenced prehistoric diplomacy?

Several factors are important in our search for the origins of diplomacy, including the emergence of tools, trade, art, gifts, and language, both spoken and written.

---

[2] Schweikard, D. P. & Schmid, H. B. (2021). Collective Intentionality. In E. N. Zalta (Ed.), *The Stanford Encyclopedia of Philosophy*. Metaphysics Research Lab, Stanford University. https://plato.stanford.edu/archives/fall2021/entries/collective-intentionality/

[3] Kurbalija, J. (2021, February 11). What can diplomacy learn from primates? Podcast Interview with Prof. Frans de Waal. https://www.diplomacy.edu/blog/interview-prof-frans-de-waal-what-can-diplomacy-learn-primates/

[4] Hobbes, T. (1996). *Leviathan*. Oxford University Press.

1. Tools

The development of tools requires a certain level of cognitive ability, cooperation, and creativity. Approximately 1.5 million years ago, our ancestors started mastering the technology of fire.[5] Cooking food led to better nutrition. Less time was spent chewing and eating, and being able to start a fire anywhere increased the mobility of humans. Making and controlling fire was a crucial stage in what would eventually become the development of all other technologies, from creating ceramic and metal goods to building today's nuclear industries.

The first tools were made of stone, such as those of the Oldowan stone type found in Tanzania, and were used from 2.3 to 1.4 million years ago.[6] Humans further mastered the art of shaping stones and began creating more sophisticated tools, such as those of the Acheulean type.[7] Humans started using more refined materials, such as the sharp edges of obsidian, around 700,000 BC. In the past 100,000 years, the development and use of tools have accelerated significantly.

An Acheulean Handaxe
Photo courtesy of Descouens, Didier. An Acheulean Handaxe, Haute-Garonne France. Lower Paleolithic - Acheulan. Chert, Muséum de Toulouse.

Making and using tools requires imagination, planning, and abilities that go beyond just finding things in nature and using them. These new cognitive skills didn't just give us new tools, but most probably coincided with the appearance of the first languages.

---

[5] Gowlett, J. A. J. (2016). The discovery of fire by humans: A long and convoluted process. *Philosophical Transactions of the Royal Society B: Biological Sciences, 371*(1696). https://doi.org/10.1098/rstb.2015.0164

[6] Wikipedia. (n.d.). Oldowan. https://en.wikipedia.org/wiki/Acheulean

[7] Wikipedia. (n.d.). Acheulean. https://en.wikipedia.org/wiki/oldowan

PREHISTORY: THE EARLY ORIGINS OF DIPLOMACY AND 'TECHNOLOGIES'

2. Trade

Another important factor in the development of diplomacy was the emergence of trade. Both activities – trade and diplomacy – require engaging others in negotiations and building trust. Archaeological evidence of trade has been found at various sites around the world. The earliest example, dating back to 300,000 BC, comes from Kenya, where proto-crayons were found.[8] The pigments in these crayons couldn't be sourced locally and must have been imported, which suggests an element of exchange. The use of pigments is also interesting in how humans began moving beyond purely useful items. They wanted to beautify their objects and began thinking about aesthetics, another sign of their new cognitive abilities.

In Europe, the River Danube played an extremely important role from 35,000 BC to 8,000 BC, in connecting western and eastern Europe. It was one of the most used long-distance water routes (going from present-day Germany to Romania) before the Mediterranean Sea became the main trade link between Europe's eastern and western regions.[9] The Vinča settlement in present-day Serbia provides evidence that goods were imported from very far away.[10] This commercial phenomenon did not exclusively happen in Europe. In Asia, we have archaeological evidence of jade trading.[11] At that time, the Indian Ocean was probably used as the major trading route between Africa and India.[12] In the Americas, researchers found traces of cocoa in jars in Central America that had to be imported from South America.[13]

These archaeological records all indicate that trade negotiations and encounters took place alongside the development of more complex forms of diplomacy.

---

[8]   Boissoneault, L. (2018, March 15). Colored pigments and complex tools suggest Humans were trading 100,000 years earlier than previously believed. *Smithsonian Magazine*. https://www.smithsonianmag.coture/colored-pigments-and-complex-tools-suggest-human-trade-100000-years-earlier-previously-believed-180968499/.m/science-na

[9]   Séfériadès, M. L. (2010). Spondylus and long-distance trade in Prehistoric Europe. In D. W. Anthony & J. Chi (Eds.). *The Lost World of Old Europe: The Danube Valley, 5000-3500 BC* (pp.179–89). Princeton University Press.

[10] Wachter, K. (2013). Uncovering Prehistoric Danube Culture | ICPDR – International Commission for the Protection of the Danube River. *Danube Watch* 2. https://www.icpdr.org/main/publications/uncovering-prehistoric-danube-culture

[11] Hung, H.-C. et al. (2007). Ancient jades map 3,000 years of prehistoric exchange in Southeast Asia. *Proceedings of the National Academy of Sciences, 104*(50), 19745–50. https://doi.org/10.1073/pnas.0707304104

[12] Seland, E. H. (2014). Archaeology of trade in the Western Indian Ocean, 300 BC–AD 700. *Journal of Archaeological Research, 22*(4), 367–402. https://doi.org/10.1007/s10814-014-9075-7

[13] Dell'Amore, C. (2011, March 29). Prehistoric Americans traded chocolate for turquoise? *National Geographic News*. https://www.nationalgeographic.com/history/article/110329-chocolate-turquoise-trade-prehistoric-peoples-archaeology

3. Art

The presence of art implied a few developments that could have set the stage for proto-diplomacy: a cognitive capacity for abstract thinking, self-awareness (evident through the recognition of one's surroundings), and an understanding of the group in which the artists lived.

From 40,000 BC to 35,000 BC, a cognitive revolution happened that led to behavioural modernity, in other words, the birth of abstract thinking, planning depth, symbolic behaviour (e.g. art, ornamentation), music, and dance.[14]

The earliest examples of art were in the form of cave paintings, including the oldest animal painting from Sulawesi, Indonesia (44,000 BC)[15]; animal and hand paintings and engravings from Altamira, Spain (36,000 BC)[16]; and close to 600 paintings found in Lascaux, France (17,000 BC).[17] Prehistoric anthropomorphic and zoomorphic sculptures came later, and include the Lion Man of the Hohlenstein-Stadel, Germany (around 37,000 BC)[18]; the earliest image of a human being, Venus of Hohle Fels, found in present-day Germany (37,000 BC)[19]; and Venus of Willendorf, Austria (25,000 BC).[20]

Art and culture have been transmitted across time and space. Diplomacy has always been, and remains, one of the main conveyor belts of artistic and cultural cross-fertilisation.

4. Gifts

Early forms of diplomacy included the exchange of gifts between groups and tribes.[21] The research of anthropologists Bronisław Malinowski and Marcel Mauss has well documented this social phenomenon. Through gifts, our distant predecessors developed links with their neighbours and other tribes. Gifts were not only transactional; their exchange developed trust and established contacts that could be useful in times of conflict. Gifts remain an important lubricator of diplomacy today.

---

[14] Wurz, S. (2012). The transition to modern behavior. *Nature Education Knowledge*, *3*(10). https://www.nature.com/scitable/knowledge/library/the-transition-to-modern-behavior-86614339/

[15] Brumm, A., et al. (2021). Oldest cave art found in Sulawesi. *Science Advances*, *7*(3), eabd4648. https://doi.org/10.1126/sciadv.abd4648

[16] Saura Ramos, P. A. (1999). *Cave of Altamira*. Abrams, Inc.

[17] Leroi-Gourhan, A. (1962). The archaeology of Lascaux Cave. *Scientific American*, *246*(6), 104–13.

[18] Ulm Museum. (2013). *The Return of the Lion Man: History - Myth - Magic*. Thorbecke.

[19] Dixson, A. F., & Dixson, B. J. (2011). Venus figurines of the European Paleolithic: Symbols of fertility or attractiveness? *Journal of Anthropology*, 1–11.

[20] Antl-Weiser, W. (2009). The time of the Willendorf figurines and new results of Palaeolithic research in Lower Austria. *Anthropologie (1962-)* 47(1/2), 131–41.

[21] See the wider discussion on gift-giving, including in the Kula Ring, in Wilton S. Dillon, *Gifts and Nations: The Obligation to Give, Receive and Repay* (Routledge, 2017).

PREHISTORY: THE EARLY ORIGINS OF DIPLOMACY AND 'TECHNOLOGIES'

Gobierno de Cantabria, Altamira Cave Painting.

Gifts were often exchanged as part of intermarriages among members of different groups. Our distant predecessors realised, very early on, the risks of inbreeding and started avoiding it by looking for mates in other clans, tribes, and groups; this guaranteed genetic diversity.[22] Through intermarriage, intertribal bonds and alliances were created. This has been a basic and enduring diplomatic practice.

> Intermarriages were particularly important in the Middle Ages, as shown by Hans Holbein's iconic painting *The Ambassadors* (1533), which depicts two diplomats negotiating the annulment of the marriage between England's King Henry VIII and Spain's Catherine of Aragon. In 1534, the failure of these negotiations triggered the separation of England from the Roman Catholic Church and its embrace of the Protestant Reformation.

5. Language
Our ability to communicate via spoken and written language makes us a unique biological species. Speech helps humans share their feelings and thoughts with others, transmit knowledge, and cement social links. In turn, understanding others led to more empathy, trust, and peaceful solutions to conflicts, collectively known as proto-diplomacy.

One of the earliest writing records can be found on the Ishango bone (20,000 BC), found in present-day Congo. The markings on the bone are interpreted as numerals,

---

[22] Martin Sikora et al., 'Ancient Genomes Show Social and Reproductive Behavior of Early Upper Paleolithic Foragers', *Science* 358, no. 6363 (3 November 2017): 659–62, https://doi.org/10.1126/science.aao1807.

HISTORY OF DIPLOMACY AND TECHNOLOGY

Joeykentin. Bone tool and possible mathematical device that dates to the Upper Paleolithic Era discovered in Ishango. Own work.

but several theories exist regarding what these etchings represent, including a counting tool, a lunar calendar, and a numeric reference table. Researchers agree on one thing: The markings are not random and are likely evidence of prehistoric numerals.[23]

In the next chapter, following this discussion of diplomacy and technology in prehistory, we turn to the Ancient world and its varied developments in diplomacy and technology, including multiple Ancient civilisations and the development of written communications.

---

[23] Gheverghese. J. G. (2011). *The crest of the Peacock: Non-European roots of mathematics, 3rd edition.* Princeton University Press.

2. Ancient Era: Cradle of humanity and diplomacy

The Ancient Era is the cradle of humanity and diplomacy. It is described as the Axial Age, when humans were put at the centre of philosophies and religions. Civilisational achievements in diplomacy developed all over the world concurrently. Besides the Mediterranean and the Middle East, there are examples of diplomacy and advanced Ancient civilisations in China, India, Africa, and the Americas.

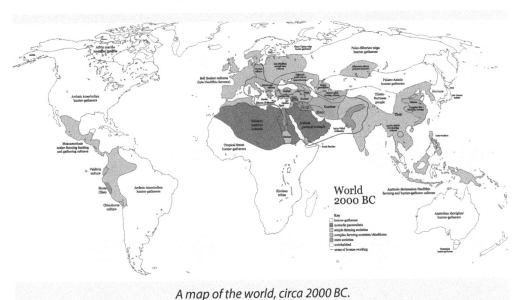

A map of the world, circa 2000 BC.

Around the 4th millennium BC, in the region of the Fertile Crescent, covering Mesopotamia (present-day Iraq, Syria, and South-Eastern Turkey) and the East Mediterranean, hunter-gatherer society was slowly making room for a new period. This phase is distinguished by new activities, such as humans cultivating plants, breeding animals for food, and forming permanent (sedentary) settlements, which would later develop into the first states.

In this period, along with agriculture and the domestication of animals, writing emerged as a key way of conveying knowledge, conserving human experience, and developing diplomacy.

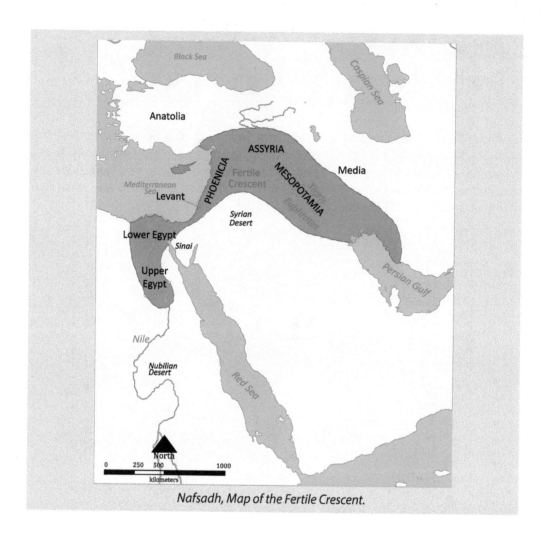

Nafsadh, Map of the Fertile Crescent.

Writing as the key diplomatic technology

Writing was, is, and will remain the key diplomatic technology. Describing writing as a technology might sound counterintuitive. Although writing is deeply integrated into the way we live and work, it is not a natural faculty like walking and speaking. Writing was invented and required tools and skills. These are the defining elements of technology. Like all other technologies, writing has shaped our way of life.

The transition from prehistory to Ancient history can be traced back to the 4th millennium BC, when the Sumerians invented writing. Archaeological evidence for this has been discovered in the form of clay tablets containing the first texts written in Cuneiform. Around 3200 BC, writing began in ancient Egypt. By 1300 BC, a fully operational writing system was used in the late Shang dynasty in China.

ANCIENT ERA: CRADLE OF HUMANITY AND DIPLOMACY

Between 900 BC and 600 BC, writing also appeared in the cultures of Mesoamerica, a historical region in North America.

Thoth, the Ibis-headed God of Books, of Writing and of all Science the Chief Friend of Man After Death.

Mesopotamian diplomacy

The Sumerians, the early inhabitants of Mesopotamia, invented writing sometime in the fourth millennium BC. But before writing was developed sufficiently to convey words accurately, couriers had to memorise and deliver messages to their recipients. The poem *Enmerkar and the Lord of Aratta*, written in Sumerian around 2750 BC, speaks about the role of messengers at the time.[24]

The extinct Akkadian language was the first diplomatic language. It was a form of lingua franca and the international tongue of the Middle East until replaced by Aramaic. Archaeologists discovered the first written diplomatic documents on clay tablets using cuneiform characters that date back to 2500 BC. An important diplomatic correspondence from circa 1300 BC (the treaty between Egypt and the Hittites) was written in Akkadian (Babylonian). The

---

[24] Woods, C. (2010). The earliest Mesopotamian writing. In C. Woods (Ed.), *Visible language: Inventions of writing in the Ancient Middle East and beyond* (pp. 33–50). Oriental Institute of the University of Chicago.

Akkadian term *mar shipri*, which first appeared in texts at the end of the third millennium BC, may designate a 'messenger', 'envoy', 'agent', 'deputy', 'ambassador', or 'diplomat'.[25]

In the Babylonian Era, during the rule of Hammurabi (18th century BC), a highly functional system of messengers was developed. According to Jack M. Sasson in his book *From the Mari Archives: An Anthology of Old Babylonian Letters*, in the same period, there was a well-developed system of envoys ranging from simple messengers to 'plenipotential ambassadors' empowered to negotiate agreements on behalf of their masters.[26] The Mari archives also include the first references to diplomatic immunities, diplomatic passports, and letters of accreditation.

Louvre Museum, Stele of Hammurabi, circa 1793-1751 BC.

Hammurabi is best known for issuing the *Code of Hammurabi*, the first legal code. If we remove the harsh punishments that the code had, we can find diplomatic techniques

---

[25] Seri, A. (2010). Adaptation of cuneiform to write Akkadian. In C. Woods (Ed.), *Visible language: Inventions of writing in the Ancient Middle East and beyond* (pp. 85–93). Oriental Institute of the University of Chicago.

[26] Sasson, J. M. (2017). *From the Mari archives: An anthology of Old Babylonian letters*. Pennsylvania State University Press.

ANCIENT ERA: CRADLE OF HUMANITY AND DIPLOMACY

that are missing today. Hammurabi understood the phrase 'skin in the game' quite literally and believed it to be a very important legal concept: If builders built a building that later collapsed and killed people, they also needed to be killed. In this way, he held them existentially accountable. He paid attention to the importance of aligning incentives, managing risk, and, most importantly, communicating legal rules and standards in a simple and understandable language to his citizens.[27]

Ancient Egypt: Amarna diplomacy

Three centuries after Ancient Babylon, Amarna diplomacy emerged. It is usually singled out as the most developed diplomatic system in Ancient civilisations, comprising the main diplomatic techniques, including the sending of representatives, negotiating, and the handing out of immunities.[28]

Amarna diplomacy is named after the Egyptian city of Tel-el Amarna, where archaeologists discovered the first diplomatic archive – the Amarna Letters. Tel-el Amarna was established by Akhenaten and served as the new capital of Egypt until the end of the 18th dynasty. Akhenaten was a leader inclined to change. He began a religious revolution in which he declared Aten a supreme god, turning his back on the old traditions.

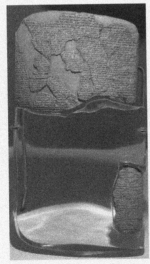

*Museum of the Ancient Orient,
Table of Treaty of Kadesh from Boğazköy, Turkey.*

---

[27] Van De Mieroop, M. (2008). *King Hammurabi of Babylon: A biography.* John Wiley & Sons Ltd.
[28] Cohen, R., & Westbrook, R. (2002). *Amarna diplomacy: The beginnings of International Relations.* JHU Press.

The dynasties of the New Kingdom of Egypt oversaw a period of extensive creativity, particularly noticeable in architecture. Additionally, diplomacy was favoured over war.

During the reign of Ramses II, the first peace treaty, the Treaty of Kadesh, was signed. After the Battle of Kadesh, considered today to have been a draw, both the Egyptian and Hittite sides decided to end hostilities. The two kings, Ramses II and Hattusili III, realised that neither could gain a substantial advantage over the other and that the best course was the path of peace. The Hittites and the Egyptians then entered into a new relationship in which they shared their knowledge and experiences. The Hittites taught the Egyptians how to make superior weapons and tools, while the Egyptians, masters of agriculture, shared their own knowledge with the Hittites. The two civilisations continued a mutually beneficial relationship until the fall of the Hittite Empire c. 1200 BC.[29]

Amarna Letters

The Amarna Letters, discovered in 1887, present an archive written on 382 clay tablets, primarily consisting of diplomatic correspondence between the Egyptian administration and the leaders of neighbouring kingdoms.

Osama Shukir Muhammed Amin, Five Amarna letters on display (G57/dc8) at the British Museum, London.

The tablets cover the reigns of Amenhotep III, Akhenaten, and possibly Smenkhkare or Tutankhamun of the 18th dynasty of Egypt. The system continued to be used for approximately 100 years after the end of the Amarna Period. The letters were written in

---

[29] Bell, L. (2008). Conflict and reconciliation in the Ancient Middle East: The clash of Egyptian and Hittite chariots in Syria, and the world's first peace treaty between "Superpowers". In K. A. Raaflaub (Ed.), *War and Peace in the Ancient world* (pp. 98–120). John Wiley & Sons Ltd.

Akkadian, which helped facilitate foreign correspondence by filtering out inappropriate language. This aided the relative peace of the time. The letters present the earliest examples of international diplomacy, while their most common subjects are negotiations of diplomatic marriages, friendship statements, and exchanged materials.

Ancient Assyrian diplomacy

As the Egyptian and Hittite Empires weakened, the Assyrian state emerged and reached its zenith during the era of the Sargonid dynasty (7th and 8th century BC), and the reigns of Sargon, Sennacherib, and Ashurbanipal. The Assyrian state tried to extend its control over the historical Fertile Crescent. They were particularly interested in gaining control over the key trade routes. The powerful Assyrian neighbours were forced to form coalitions and alliances to counterbalance this emerging Assyrian power. Assyrian dynasties used both war and diplomacy to achieve their goals.

Library of Ashurbanipal Mesopotamia 1500-539 BC Gallery, British Museum, London.

The main Assyrian archaeological source is the Library of Ashurbanipal, which was discovered during excavations of the imperial palaces in Nineveh and Kuyunjik. The Library, consisting of clay tablets, is a rich source of materials, covering all facets of both social and official Assyrian life, including diplomacy and communication. More specific details about their communication systems were also discovered, including the existence of a beacon scheme (an early telegraph system). It seems that this scheme was used for transmitting messages.

Ancient Persian diplomacy

The Persian Empire was hegemonic, relying more on military might than on diplomatic skills. Limited forms of diplomatic interaction were developed with the city-states of Ancient Greece. One of the main legacies of the Persian Era was a highly developed communication system – the Ancient-Era internet.

The Greek historian Xenophon described the messaging system in Persia during the rule of King Cyrus the Great (599 BC to 530 BC) as follows:

> We have observed still another device of Cyrus to cope with the magnitude of his empire; by means of this institution he would speedily discover the condition of affairs, no matter how far distant they might be from him: he experimented to find out how great a distance a horse could cover in a day when ridden hard, but so as not to break down, and then he erected post-stations at just such distances and equipped them with horses, and men to take care of them; at each one of the stations he had the proper official appointed to receive the letters that were delivered and to forward them on, to take in the exhausted horses and riders and send on fresh ones. They say, moreover, that sometimes this express does not stop all night, but the night messengers succeed the day messengers in relays, and when that is the case, this express, some say, gets over the ground faster than the cranes.[30]

Although the messenger system was developed for military purposes, it began to be used for receiving and sending envoys to neighbouring states and tribes. The Persian Empire discovered the potential of diplomacy in the last days of its existence when Darius III offered peace to Alexander of Macedonia based on 'ancient friendship and alliance'. The refusal of this offer led to the conquering of the Persian Empire by the Greeks and the end of the period of Ancient civilisations.

Ancient Chinese diplomacy

The earliest records of Chinese diplomacy date from the first millennium BC. By the 8th century BC, the Chinese had leagues, missions, and an organised system of polite dialogue between their many feuding Chinese kingdoms, including resident envoys who served as hostages to the good behaviour of those who sent them. The sophistication of this tradition, which emphasised the practical virtues of ethical behaviour in relations between states (no doubt in reaction to actual amorality), is well documented

---

[30] Xenophon. (trans. 1961). *Cyropaedia* (p. 419), trans. Walter Miller. William Heinemann.

in the Chinese classics. Its essence is perhaps best captured by the advice of Zhuangzi to 'diplomats' at the beginning of the 3rd century BC. He advised them thus:

> If relations between states are close, they may establish mutual trust through daily interaction; if relations are distant, mutual confidence can only be established through exchanges of messages. Messengers (diplomats) are required to deliver messages. Their contents may be either pleasing to both sides or likely to engender anger between them.
>
> Faithfully conveying such messages is the most difficult task under the heavens, for if the words are such as to evoke a positive response on both sides, there will be the temptation to exaggerate them with flattery, and if they are unpleasant, there will be a tendency to make them even more biting. In either case, the truth will be lost. If the truth is lost, mutual trust will also be lost. If mutual trust is lost, the messenger himself may be imperilled. Therefore, I say to you that it is a wise rule: "always speak the truth and never embellish it. In this way, you will avoid much harm to yourselves."[31]

Chinese philosophers thought that the best way for a state to exercise influence abroad was 'to develop a moral society worthy of emulation by admiring foreigners and waiting confidently for them to come to China to learn'.[32]

Ancient Indian diplomacy

Kautilya's book *Arthashastra*, one of the earliest works of Sanskrit literature, preserves the history of Ancient India's diplomatic tradition. The ruthless state system graded state power based on five factors and emphasised espionage, diplomatic manoeuvre, and contention by 12 categories of states within a complex geopolitical matrix. It also proposed four means of statecraft (conciliation, seduction, subversion, and coercion) and six forms of state policy (peace, war, nonalignment, alliances, shows of force, and double-dealing). Ancient India fielded three categories of diplomats – plenipotentiaries, envoys entrusted with a single issue or mission, and royal messengers, a type of consular agent in charge of managing commercial relations and transactions, and two kinds of spies – those charged with the collection of intelligence and those entrusted with subversion and other forms of covert action.

---

[31] Britannica. (n.d.). Diplomacy - History of diplomacy. https://www.britannica.com/topic/diplomacy/History-of-diplomacy

[32] Britannica. (n.d.). Diplomacy - History of diplomacy. https://www.britannica.com/topic/diplomacy/History-of-diplomacy

Arthashastra also describes the detailed rules that regulate diplomatic immunities and privileges, the inauguration and termination of diplomatic missions, and the selection and duties of envoys. Thus, Kautilya describes the 'duties of an envoy' as:

> 'sending information to his king, ensuring maintenance of the terms of a treaty, upholding his king's honour, acquiring allies, instigating dissension among the friends of his enemy, conveying secret agents and troops [into enemy territory], suborning the kinsmen of the enemy to his own king's side, acquiring clandestinely gems and other valuable material for his own king, ascertaining secret information, and showing valour in liberating hostages [held by the enemy]'.[33]

He further stipulates that no envoys should ever be harmed, and even if they deliver an 'unpleasant' message, they should not be detained.

The Indian subcontinent, isolated from its neighbours by deserts, seas, and the Himalayas, had very little political connection with the affairs of other regions of the world until Alexander the Great conquered its northern regions in 326 BC. Later on, the establishment of the native Mauryan Empire opened the way for a new era in Indian diplomatic history.

The Mauryan emperor Ashoka received a few emissaries from the Macedonian kingdoms. The Indian way of thinking and its early traditions of diplomatic reasoning were lost and replaced by those of its Muslim and British conquerors. The exception was South India, where the Chola dynasty and other kingdoms continued diplomatic and cultural exchanges with Southeast Asia and China and preserved the text and memories of the *Arthashastra*.

From this overview of diplomacy and technology in Ancient civilisations, in which we saw the development of written communication and the use of clay tablets to transmit information, we move on to focus on one prominent Ancient civilisation: Ancient Greece. This period enjoyed several technological developments which were crucial for the diplomacy of the time.

---

[33] Kauṭalya. (1992). *The Arthashastra* (p. 576). Penguin Books India.

3. Ancient Greece: Politics, negotiations, and diplomacy

The most direct bridge between Ancient civilisations and our era is Ancient Greece, often considered as the cradle of modern civilisation. Greek notions of rational thought, argumentation, and competition are the foundation of the way we think today. As the renowned British philosopher A.N. Whitehead once commented, the European philosophical tradition 'consists of a series of footnotes to Plato'.[34] Modern sciences also owe much to Ancient Greece. It was in Greece that philosophers and scientists developed the method of testing ideas against each other. In Greek politics, democracy was developed through a battle of ideas and exercising power, to gain the trust and support

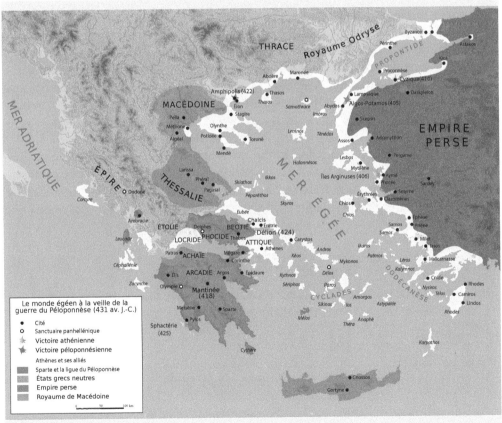

Marsyas. Carte du monde égéen en 431 av. J.-C., à la veille de la Guerre du Péloponnèse.

---

[34] North Whitehead, A. (1978). *Process and reality* (p. 39). The Free Press.

of those who voted. Theatre, geometry, and astronomy also originated in Greece, as did the terms 'diplomacy' and 'technology'.

Etymology of diplomacy and technology

The term diplomacy derived from the Ancient Greek *diplōma* (δίπλωμα) composed of *diplo*, meaning 'folded in two', and the suffix *-ma*, meaning 'an object'. The folded document granted a privilege to the bearer, often a permit to travel. Later, the term *diploma* was borrowed by the Latin language to mean an official document. In the 18th century, the French term *diplomate* referred to a person authorised to negotiate on behalf of a state, and *corps diplomatique* referred to officials involved in foreign policy. In 1796, the term diplomacy was officially introduced to the English language by philosopher and political scientist Edmund Burke.

The term technology is derived from the Ancient Greek word *technelogia* (τεχνολογία), which is a mix of *techne* (art, method, tool, skill, craft) and the suffix *-logia* (knowledge), translating as the 'science of craft'. Aristotle defined *techne* as a 'rational faculty exercised in making something … a productive quality exercised in combo with true reason'.[35] He believed that the business of techne is to 'bring something into existence which has its efficient cause in the maker and not in itself'. It is also important to note that Aristotle related *techne* to the crafts and sciences, most notably through mathematics. Among historians, there is an ongoing debate about whether Ancient Greece was an advancement or regression in the development of diplomacy compared to earlier practices present in Ancient Egypt and Persia. Diplomacy Historian, Professor Raymond Cohen, argues that 'the practice of Greek diplomacy was quite rudimentary' and 'compared with Persian cosmopolitanism, Greek diplomacy was provincial and unpolished'.[36]

Early techno-scepticism

The invention of writing also triggered the first criticism of its impact on society. In Plato's *Phaedrus*[37] Socrates narrates how Theuth (Thoth), the Egyptian god of inventions, tried to persuade the sceptical Egyptian king Thamus (Ammon) about the value of the new invention – writing. In the process, Socrates faulted writing for weakening the necessity and power of memory and for allowing the 'pretence of understanding', rather than 'true understanding'. Socrates' Thamus argued that:

---

[35] Aristotle, *Nicomachean Ethics*, second edition, translated by Terence Irwin, Indianapolis: Hackett Publishing Co., 1999.
[36] Cohen, R. (2018). Diplomacy through the ages. In P. Kerr & G Wiseman (Eds.), *Diplomacy in a Globalizing World: Theories and Practices* (pp. 15-20). Oxford University Press.
[37] Plato. (2005). *Phaedrus*. Penguin Publishing Group.

ANCIENT GREECE: POLITICS, NEGOTIATIONS, AND DIPLOMACY

> 'For this invention will produce forgetfulness in the minds of those who learn to use it, because they will not practice their memory. Their trust in writing, produced by external characters which are no part of themselves, will discourage the use of their own memory within them.'

Geopolitical context for Greek diplomacy

Located in Southeast Europe, Ancient Greece was the bridge between Asia and Europe. This fact is important for understanding the overall interplay between diplomacy and war. Greece also had over 1,000 city-states. These states were not only independent but also interdependent, which created fertile ground for diplomacy. The balancing of power was well developed.

Mythology, Oratory, and Writing

The god Hermes was the protector of traders. His task was to negotiate between warring parties and other gods. According to British diplomat and historian Harold Nicolson, Hermes had guardianship over diplomatic representatives, which had 'an unfortunate effect' on their reputation because Hermes symbolised charm, trickery,

Jastrow. So-called Logios Hermes (Hermes, Orator). Marble, a Roman copy.

and cunningness.[38] Some modern suspicion about diplomacy and diplomats could be traced back to Hermes, and his not very favourable image.[39]

Sting, Bust of the Greek orator Demosthenes. Marble, Roman artwork.

Demosthenes was one of the first ambassadors of the Ancient city-state Athens. His influence shows the importance of rhetoric in diplomacy. Sophists elevated rhetoric to the level of a creative craft. Rhetoric remains one of the key tools of diplomacy today.

Thucydides was an Athenian historian who also served as an Athenian general during the Peloponnesian War. He is also the author of *The History of the Peloponnesian War*, one of the most detailed accounts of the question of power in early international relations.

Thucydides is still part of the basic literature for the study of power and international relations. His modern relevance is attested to by Kissinger's writings on the balance of power in the 19th century, and his practice of power politics during his time as the US secretary of state.

---

[38] Kurbalija, J. (2021). *Ancient Greek diplomacy: Politics, new tools, and negotiation.* https://www.diplomacy.edu/histories/ancient-greek-diplomacy-politics-new-tools-and-negotiation

[39] Nicolson, H. (1950). *Diplomacy* (p. 19). Oxford University Press.

Shakko. Cast of a bust of Thucydides.

The diplomats of Ancient Greece

A proxenos (pl. proxenoi) was a Greek consular agent, i.e. a citizen of the city in which he resided, not of the city-state for which he worked. Like envoys, proxenoi had a secondary task of gathering information, but their primary responsibility was trade.

Proxenoi were early honorary consuls, ambassadors, and lobbyists. A proxenos would use his influence in his own city to promote policies of friendship or alliance with the city he voluntarily represented. He was expected to handle all high-level political matters, as well as provide support and housing for visitors from the sending state (merchants, representatives, politicians). Proxenoi were granted certain immunities (e.g. asylum in case the sending state turned against him, or free and safe travel during both peace and war), and the position of the proxenos held prestige and was hereditary. During the classical period, it is probable that all well-known Athenian politicians held one or more positions as proxenoi (e.g. Demosthenes was a proxenos in Athens for Thebes).

A presbus (pl. presbeis) or envoy, was a senior citizen involved in advocacy and represented a type of early public diplomat. They were great orators who went on unpaid, ad hoc missions. Delegations of envoys were often large, numbering 20–30 people. They were prominent representatives of the sending states. In many cases, they addressed the citizens and senate of the receiving state and tried to persuade the elite of the receiving state about the position of their country.

A keryx (pl. *kerykes*), was an inviolable Ancient Greek messenger. In Homer's time, the keryx was simply a trusted attendant or retainer of a chieftain. The role of kerykes expanded, however, to include acting as inviolable messengers between states, even in times of war, proclaiming meetings of the council, popular assembly, or court of law, reciting there the formulas of prayer, and summoning people to attend. They were regarded as the offspring of Hermes, the messenger of the gods. Kerykes were general-purpose couriers and masters of ceremonies. The diplomatic responsibilities of these messengers were to serve as 'truce-bearers' before the start of the Panhellenic games and to make announcements during the games. More important still was their task of going ahead of the ambassadors to secure guarantees for their safe reception. Also, they were responsible for issuing ultimatums and declarations of war. Heralds were the early masters of protocol.[40]

Innovations in diplomacy: Early multilateral diplomacy and secrecy

Early multilateral diplomacy was developed around the idea of a truce during the Olympic Games and other common festivities. At that time, the representatives of city-states used to gather. It was also a moment when they shared a common identity, and it was a good opportunity to negotiate. The idea of Common Peace was born, which meant permanent peace between the Greek city-states. With many setbacks, the idea of Common Peace could be traced to our era. It was one of the founding principles of the League of Nations and our system today, based on the United Nations (UN) Charter.

Multilateral diplomacy occurred more in the state system's religious leagues (neighbouring communities sharing a deity) and large-member military alliances (or 'leagues'), established for defence and offence. One of the examples of a well-developed multilateral alliance is the Second Athenian Confederacy, a defensive alliance created in 378/7 BC. Its purpose was to guard against a growing fear that Sparta would not honour the common peace of the Greek cities. The founding charter, the famous Decree of Aristoteles, described its purpose and defensive character and invited others to join, including any barbarians (non-Greeks) on the mainland or islands.

Dating back to the 5th century BC, steganography is one of the oldest methods of concealing secret information. According to Herodotus, it was first used by the tyrant Histius, who shaved the head of a servant before tattooing a message on his scalp. When the hair had grown back, the servant was sent to deliver his message – a warning of an impending attack by the presumably slow-moving Persian army – which was revealed when the servant's head was once again shaved.[41]

---

[40] Berridge, G. (2018). *The diplomacy of Ancient Greece - A short introduction* DiploFoundation. https://issuu.com/diplo/docs/the_diplomacy_of_ancient_greece

[41] Kahn, D. (1996). The history of steganography. In R. Anderson (Ed.), *Information hiding: First international workshop, Cambridge, UK, May 30 - June 1, 1996* (pp. 1-5). *Proceedings*. Springer.

ANCIENT GREECE: POLITICS, NEGOTIATIONS, AND DIPLOMACY

Technological breakthroughs: Early telegraphs

In the 2nd century BC, two engineers, Cleoxenes and Democletus, invented the system for conveying messages by using pyrseia (*pirsos* – torch). This system consisted of a 5x5 table containing the letters of the Greek alphabet, where each letter corresponded to a row and a column.

Pyrseia code.

The configuration of the Greek land, with its variety of mountains and hills, was convenient for conveying messages. Phryctoria were towers built on mountaintops, usually 20 miles from each other, with good visibility between them. Messages were transmitted from tower to tower, using the pyrseia table, by lighting flames on one tower and then the next tower in succession.

To send a message from one hill to the next, two groups of five torches were used. The group of torches on the left indicated the row and the group on the right specified the column of the table. The message was conveyed by identifying a specific letter through a combination of lit torches.

In the tragedy *Agamemnon*, Aeschylus describes how the message for the fall of Troy arrived at Mycenae via the use of phryctoria:

Chor. And since what time has the town been destroyed?

Clyt. Since the night, I say, that has just now given birth to the light of this morning.

Chor. And what messenger is there that could arrive with such speed as this?

Clyt. Hephaistos, sending forth from Ida a bright radiance. And beacon ever sent beacon hither by means of the courier fire: Ida (sent it) to the rock of Hermes in Lemnos; and a huge torch from the island was taken over in the third place by Zeus' peak of Athos; and paying more than what was due(?), so as to skim the back of the sea(?), the strength of the travelling torch joyously (went on ...) the pine-tree blaze, after (?) transmitting, like a sun, its golden radiance to the look-out of Makistos. And he (i.e. Makistos), not dallying nor heedlessly overcome by sleep, did not neglect his share in the messenger's duty, and afar, over the streams of Euripus, the beacon's light gave the watchers of Messapion the sign of its arrival. They kindled an answering flare and sent the tidings onward, by setting fire to a stack of aged heath. And the vigorous torch, not yet growing dim, leaped, like· the shining moon, over the plain of Asopus to the rock of Kithairon and there waked a new relay of the sender fire. And the far-sent light was not rejected by the watch-post, which burned more than it had been ordered; and the light shot down over the Gorgon-eyed lake and reaching the mountain of the roaming goats urged (the watch-post) not to neglect (?) the ordinance of the fire. And they with stintless might kindled and sent on a great beard of flame, and it passed beyond the promontory that looks down on the Saronic straits, blazing onward, and shot down when (?) it reached the Arachnaean peak, the watchpost that is neighbour to our city; and then it shot down here to the house of the Atridae, this light, the genuine offspring of its ancestor, the fire from Mount Ida. Such, thou seest, are the rules I arranged for my torch-bearers,-one from another in succession supplied to the full; and victor is he who ran first and last. Such is the proof and token that I give thee, transmitted to me by my husband from Troy.[42]

The historian Polybius later improved pyrseia and made it renowned. According to Polybius, Aeneas Tacticus invented the hydraulic telegraph around 350 BC. It was a semaphore system that was later used during the First Punic War to send messages between Sicily and Carthage.[43] How did it work? Cylindrical containers were located on stations on the selected hilltops. The containers were then filled up to the same point

---

[42] Aeschylus, *Agamemnon*, ed. Eduard Fraenkel (Oxford University Press, 1962), 109–11.

[43] Austin, N. J. E., & Rankov, N. B. (1998). *Exploratio: Military & political intelligence in the Roman World from the Second Punic War to the Battle of Adrianople*. Routledge. https://doi.org/10.4324/9780203033050.

with water. A wooden stick that carried the pre-agreed messages was mounted vertically on the cork that floated on the surface of the water. The transmitter and receiver operators had to work simultaneously to send the right message. Every time a message had to be transmitted, the sender raised a lit torch, marking the simultaneous opening of the taps. As soon as the receiver turned on its own torch, the two operators opened the tap at the same time. When the water level reached the height that corresponded to the desired message, with the same procedure as the torches, the transmitter informed the receiver to close his tap.[44]

A relief of the Greek hydraulic telegraph of Aeneas, depicting one half of a telegraph system.

---

[44] You can watch an animated demonstration of this system at https://www.youtube.com/watch?v=Az28Hsne-nI

Diplomacy and democracy: An early tension

The strong focus on openness and transparency is one of the parallels between Ancient Greece and our era. Ancient Greek diplomacy was one of the most open diplomacies ever practised. Envoys addressed public gatherings of citizens of the receiving states by using the art of rhetoric and publicising important treaties by inscribing them on stone or bronze pillars (*stelai*) located in temples or other sacred places.

Paradoxically, at least at first glance, this openness created one of the major weaknesses of Ancient Greek diplomacy. Facing the foreign public, envoys advocated rather than negotiated. Their task was to persuade the wide population of the receiving state, not to negotiate a compromise with the opposite side.

This tension between the public and the discrete nature of diplomacy still exists today. We should evaluate the open and public nature of digital diplomacy while keeping in mind the primary purpose of diplomacy: achieving peaceful solutions for conflicts through compromise. It is often difficult to reach a compromise while negotiating in the public eye. However, modern diplomacy cannot, and should not, tolerate secret diplomacy. Despite this, some form of 'translucent' diplomacy is often more effective in reaching compromise solutions than today's often choreographed public diplomacy, increasingly conducted online. This is at least one historical lesson that we can learn from Ancient Greek diplomacy.

Meanwhile...

... in China:

In 221 BC, the Qin dynasty managed to unite most of the country for the first time in 500 years, ending the period of the Warring States. Although lasting only 15 years, the first Qin emperor managed to establish a highly centralised government and develop a legal code and written language. These are just a few of the achievements, which had such a lasting effect that they gave the country its name: China.

Between 202 BC and 220 AD, the emperors of the Han dynasty governed according to Confucian principles. The Han governed for four centuries, consolidating unity in the region. This has been called China's golden period. During the Han dynasty, diplomatic relations with many different Asian states were established. All of those countries received Chinese ambassadors. Diplomacy was used to secure trade routes through regions to the west. The famous Silk Road to the Levant, over which traded goods reached the Mediterranean and Europe (at the time mostly under the Roman Empire), was established, as was the sea route to India.

Three thinkers mainly influenced the Chinese diplomacy of that period. Sun Tzu, in *The Art of War*,[45] advocated diplomacy, negotiations, and cultivating relations with other nations as essential to the benefit of the state. Even more important were the ethical teachings of Confucius, who preached respect for superior states, the observance of the legitimacy of authorities on various levels, and the distinction between 'cultured' Chinese and 'barbaric' foreigners. Confucius also advocated ritual and protocol in the highest levels of government.[46] Finally, the teachings of an earlier philosopher, Mencius, stated that the best way for a state to exercise its influence abroad is to develop a society worthy of emulation and admiration by foreigners.[47]

... in India:
In the same period, another important state to rise was the Maurya Empire in India. This empire reached its apex during the reign of Ashoka the Great (269–232 BC) who ruled over most of the subcontinent except for its southernmost part.

Ashoka was a keen Buddhist, and apart from promoting religion in his own realm, he also sent diplomatic missions to the Hellenistic kingdoms of Asia to spread the dharma (cosmic law and order), a set of political and moral ideas, though this was most probably done for the sake of expanding and promoting trade as well as political influence. Such contacts continued for the next millennia until the Rajput kingdoms (from the 8th century) again isolated North India.

Now that we have discussed diplomacy and technology in Ancient Greece, let's look at a later empire that lasted for over 1,000 years – the Byzantine Empire – in which diplomacy proved to be a key feature of the empire's longevity.

---

[45] Tzu, S. (2010) The art of war. Chichester, England: Capstone Publishing.
[46] Confucius; James Legge; Ch'u Chai; Winberg Chai. Li Chi: Book of Rites. An encyclopedia of ancient ceremonial usages, religious creeds, and social institutions, New Hyde Park, N.Y., University Books [1967]. (originally published in 1885)
[47] Britannica. (n.d.). Diplomacy - History of diplomacy. https://www.britannica.com/topic/diplomacy/History-of-diplomacy

4. Byzantine diplomacy: The elixir of strategic longevity

The term 'Byzantine' comes from the name of the Ancient Greek city Byzantium, which the Roman Emperor Constantine I (Constantine the Great) rebuilt and renamed Constantinople. Today, the city is known as Istanbul. In 330 CE, Constantine moved the capital of the Roman Empire from Rome to Constantinople. Situated on the Bosporus (part of the boundary between Europe and Asia), the city's location ensured that it could be supplied from the sea in the event of a siege.

The tale of the founding of Byzantium speaks about its favourable position. According to the legend, the Ancient Greek prince Byzas consulted the oracle of Delphi, asking where to establish a new colony. The oracle instructed him to settle opposite the 'land of the blind'. When he arrived at where the Sea of Marmara meets the Bosporus, on the border of Europe and Asia, he understood the meaning of the oracle. Opposite him was a Chalcedon colony in Asia Minor, whose inhabitants he considered 'blind' as they did not see the advantages of the European shore. In 667 BC, Byzas founded the city of Byzantium.

The Byzantine Empire (395–1204 and 1261–1453), also known as the Eastern Roman Empire, was known to its inhabitants as the Roman Empire or the Empire of the Romans, even after the fall of the Western Roman Empire in 476. During its golden era in the 6th century under Emperor Justinian, this empire covered much of the land surrounding the Mediterranean Sea, including what is now Italy, Greece, and Turkey, along with portions of North Africa and the Middle East.

One of the most impressive achievements of the Byzantine Empire was its longevity (330–1453), and it remains one of the longest-lasting governmental organisations over the last two millennia. The empire survived through adaptation; its backbone was its administration, which managed to adapt to the frequent changes of rulers and continuous crises on its borders.

With a relatively limited military force, diplomacy was crucial for its survival. Diplomacy was necessary because its limited military strength could not protect its long borders. Thus, one of the golden rules of the Byzantine elite was to avoid war at all costs. Byzantine rulers were aware that, even with an occasional victory, the empire would lose in the long term if it engaged in military conflicts. Wars were expensive even then and cost much more than bribing the enemy or resolving a dispute.

Byzantine diplomacy: Inherited practices and innovations

Byzantine diplomacy was the bridge between Ancient and modern diplomacy. The empire was both a practitioner of the diplomatic traditions of Ancient Mesopotamia, Egypt, Greece, and Rome, as well as an innovator of new practices that have been passed on to our time via Venice and the diplomacy of Renaissance Italy.

Byzantine diplomacy inherited and applied the following practices from Ancient civilisations:

- From the Hellenistic East (Mesopotamia, Egypt, Persia): elaborate protocol and ceremonies, dynastic marriages to cement an alliance, and trade diplomacy by merchant ambassadors.

- From Ancient Greece: the use of rhetoric as a tool of public diplomacy (although Byzantine envoys relied far less on oratory than those dispatched by the Greek city-states).

- From the Roman Empire: divide and conquer tactics, and using civil-engineering projects to impress foreigners

A proto-ministry of foreign affairs

In the Byzantine Empire, diplomacy evolved from ad hoc to organised government activities through its Office of Barbarian Affairs. The term 'barbarian' referred to all foreigners, thus, the office can be understood as an 'office of foreign affairs'. The office housed numerous interpreters and translators, and similar to a modern ministry of foreign affairs, it prepared Byzantine envoys for missions abroad, analysed reports that arrived from envoys, organised visits of foreign dignitaries to Constantinople, and prepared international treaties, among other things. The office established archives as a way to preserve institutional memory. Unlike modern diplomacy, the Byzantine Empire did not have permanent diplomatic missions. Envoys were sent abroad to deal with specific issues, even when they took several years to solve.

Soft power, public diplomacy, and protocol

The Byzantines realised the importance of winning the hearts and minds of their neighbours early on. Whenever possible, the empire transformed neighbours from potential enemies into allies and friends. Their public diplomacy toolkit included the conversion of nomadic tribes (especially Slavs) to Christianity, the use of ceremonies for impressing foreign dignitaries, and the education of future neighbouring rulers in Constantinople's leading schools. Some of these public diplomacy techniques are still in use today.

The empire particularly excelled in the use of protocol and ceremonies to impress foreign dignitaries. The location and civil engineering of Constantinople (with buildings like the Hagia Sophia and the Hippodrome of Constantinople) remain impressive even today.

The Byzantine emperor hosted foreigners in the Magnaura, where he was seated on an automated golden throne, surrounded by golden animals like roaring lions and twittering birds.[48,49] A special hydraulic system elevated the throne to the ceiling, making a lasting impression on visitors. The *Book of Ceremonies*, compiled by Constantine VII (945–959), describes court rituals, seating arrangements, and other protocol details.[50]

Introduction of regular diplomatic reporting

Envoys had an obligation to send written diplomatic reports back to Constantinople. The reports were archived in the Office of Barbarian Affairs. The content of the diplomatic reports was not very different from modern diplomatic reports, as we could see from WikiLeaks' diplomatic reports and cables. They dealt with things like local political developments, the personalities of leaders, and analyses of power struggles.

Proto intelligence service

Access to information was central to Byzantine diplomacy. To obtain the right information, the empire founded the first intelligence service, consisting of a network of official and unofficial agents (including merchants, missionaries, and military officers) who were sent abroad. To ensure secure communication, they further enhanced the Caesar cypher encryption technique.

---

[48] Filson, L. (n.d.). Special Topics Lecture 4: Byzantine technology. https://filsonarthistory.wordpress.com/2019/01/16/special-topics-lecture-4-byzantine-technology/ See also: Gerard Brett, "

[49] Brett, G. (1954). The automata in the Byzantine "Throne of Solomon". *Speculum, 29*(3).

[50] Constantine Porphyrogennetos. (2017). *The Book of Ceremonies*, trans. A. Moffatt * M. Tall. Brill.

Early multistakeholder diplomacy

The Byzantine Empire used the services of merchants, priests, and other citizens who travelled abroad. All of them served as Byzantine diplomats and had a duty to report back to Constantinople from their travels. In this way, Byzantine diplomacy managed to achieve the difficult task of maintaining a huge empire with very limited military power and financial resources.

Early international law

The Byzantines signed international treaties with neighbouring tribes, although these were framed as unilateral decrees since the emperor claimed to be the ruler of the whole world.

In the long term, the empire benefited from establishing predictable and legal relations with otherwise unruly tribes on its borders. Even when it had to go to war, it tried to find legal justifications such as the concept of 'just war' (reclaiming lost territories or defending the empire). The introduction of a legal aspect to relations with foreigners had an important impact on developing more civilised relations in Europe and the Mediterranean.

In some cases, adherence to legality restricted room for manoeuvring for the otherwise very pragmatic Byzantines, but even in these situations, they avoided breaking the rules. They employed a very complex interpretation of treaties to justify their actions in accordance with signed treaties, making them masters of interpretation and constructive ambiguity.

Mastering time management

Time always played an important role in the Byzantine Empire. During its 1,123 years of existence, the empire preserved many distinctive features, including how it conducted diplomacy. Byzantine diplomacy mastered the use of time in its activities. In most cases, playing the waiting game was in the Byzantine Empire's best interest. With stable institutions, Byzantines were always ahead of nomadic tribes, which were easily affected by disease and the weather. Thus, in times of conflict, the smart approach was to let time pass to defuse tension and choose the right moment for counter-action.

Byzantine communication and technology

Byzantine contributions to technology, philosophy, science, art, and architecture are numerous and include both inventions and innovations.[51] Communication across the expanses of the empire was necessary for both diplomacy and administration. During the 9th century, the Byzantines used a system of beacons during the Arab–Byzantine wars to transmit messages from the border with the Arab caliphate, across Asia Minor, to Constantinople. The system was devised by Leo the Mathematician during the reign of Emperor Theophilos (829–842 CE). The main line of beacons stretched over some 450 miles and functioned through two identical water clocks placed at two terminal stations.

From Byzantine to modern diplomacy

Many core elements of Byzantine diplomacy exist even today. They were passed on to the modern era via Italian city-states – mainly Venice. For example, Italian city-states established the first permanent embassies. Newly developing European nation states borrowed Italy's diplomatic style. France created the first formal ministry of foreign affairs in the 17th century, and other European states followed.

A few lessons from Byzantine diplomacy remain relevant for our time as well:

- Time management: Successful diplomacy requires consistent efforts over a long period. Diplomacy requires time, which opposes our continuous need for immediate action.

- Centrality of information: The Byzantines realised the importance of knowing what was happening abroad, and they used this information to control neighbouring tribes for over 1,000 years. The advancement of technology, and especially the internet, has brought a significant asymmetry between players in the policy field.

- Gradual innovation: The Byzantines learned new techniques, collected feedback, and then included it in their revisions. They were masters of the Gartner hype cycle,[52] and understood the phases of emerging ideas.

---

[51] McClellan, J. E., & Dorn, H. (2006). *Science and technology in world history*. The Johns Hopkins University Press.
[52] Wikipedia. (n.d.). Gartner hype cycle. https://en.wikipedia.org/wiki/Gartner_hype_cycle

- Don't forget the basics: We must always remember the purpose of diplomatic work. In the case of the Byzantines, it was the survival of the empire. Despite numerous difficulties, they managed to last over 1,000 years.

Today, the adjective 'byzantine' is sometimes used to describe a devious and usually secretive manner of operation: intrigue, plotting, and bribing. Still, historical records show that Byzantine politics were morally neither worse nor better than politics in previous or later years. In the end, their diplomacy served the Byzantines well.

Meanwhile…

…in Asia:

In the 13th century, the largest contiguous land empire in history was growing in East Asia: the Mongol Empire. The Mongol Empire emerged from the unification of nomadic tribes in present-day Mongolia under the leadership of Genghis Khan, who ruled from AD 1206 to AD1227. Setting out on numerous ferocious military campaigns, he spread his rule from Korea to the Caspian Sea, while his successors continued the expansion, reaching Poland and Asia Minor in the west and conquering the whole of China.

A major Mongol contribution was building trade routes across Eurasia, including the famous Silk Road, on which Marco Polo travelled. The routes were safe and well maintained, ushering in an era of great exchange between the West and the East.

The Mongols were open to various influences and practised religious tolerance, merit over birth, and equality before the law. Their approach to diplomacy was also open: In contact with other nations, the empire employed specialised personnel who knew how to adapt to the cultural frameworks of others. Mongol diplomacy aimed to set up a network of more or less formal dependencies. In short, the rules of this vast empire were fairly simple: surrender, be loyal, and enjoy privileges – or perish.

A millennium of Byzantine diplomacy impacted later diplomatic practices. In the next chapter, we see how multiple features of Byzantine diplomacy were combined with new diplomatic and technical innovations in Renaissance Italy, to create a forerunner to diplomacy in the modern era.

5. Renaissance diplomacy: Compromise as modus vivendi

Renaissance diplomacy shaped many modern diplomacy institutions, from permanent embassies to embryonic ministries of foreign affairs. Many methods of diplomacy were mastered in relations among city-states of Renaissance Italy where diplomacy was mastered.

Societal context for renaissance diplomacy

The Renaissance (French: 'rebirth') was a period in European civilisation immediately following the Middle Ages. From the late 13th to the early 17th centuries, it brought a renewed interest in classical learning, first to Italy and later to all of western and central Europe. The following developments during this period are important for understanding how Renaissance diplomacy emerged.

Culture: Great writers, such as Francesco Petrarca, Dante Alighieri, and Giovanni Boccaccio, emerged in this period, and art reached its peak with Leonardo da Vinci, Michelangelo, and Sandro Botticelli.

J. Siebold. Waldseemüller map, 1507.

Society: European society was deeply damaged by the Black Death (the bubonic plague pandemic) in the mid-14th century. Some countries lost up to 30%–40% of their population, which brought about extreme social changes.

Church: The Great Western Schism (Great Schism) was a split within the Roman Catholic Church (1378–1417) during which there were two rival popes from Rome and Avignon, each with his own following and administrative offices. The followers of the two popes were divided chiefly along national lines, thus, the dual papacy fostered the political antagonisms of the time. These rival claims to the papal throne damaged the prestige of the Roman Catholic Church, which began to lose its authority.

Discoveries: This period was marked by major discoveries. By 1488, the Portuguese had explored and mapped the African coastline down to the Cape of Good Hope. In 1492, Christopher Columbus persuaded the king and queen of Spain to experiment with the idea of sailing west into the Atlantic when he incidentally 'discovered' the New World (North and South America).

Inventions: One of the most significant (and by far the most influential) inventions of the 15th century was the printing press, invented by Johannes Gutenberg. It was modelled on the design of existing screw presses and could produce up to 3,600 pages per workday.

Politics: When the Renaissance began in the mid-14th century in Italy, Europe was divided into hundreds of independent states, each with its own laws and customs. Monarchs in France, England, and Spain responded to the chaotic situation by consolidating their power. A significant development in all three of these monarchies was the rise of nationalism, and pride in and loyalty to one's homeland. Monarchs centralised power and replaced the small networks of principalities and duchies.

Geostrategy: During the Renaissance, the Byzantine Empire collapsed with the fall of Constantinople in 1453. Spain emerged as a new power after completing the Reconquista of the Iberian Peninsula in 1492; France united under Louis XI; Germany was divided into small principalities; Poland and Lithuania were important players in Eastern Europe; the Grand Duchy of Moscow started emerging; and in the south, the Ottoman Empire was well established in the Balkans.

Geopolitical context for Renaissance diplomacy

The most developed form of Renaissance diplomacy appeared among Italian city-states in the 15th century, which is considered to be the beginning of modern diplomacy as we know it today. It included permanent diplomatic missions and a rudimentary ministry of foreign affairs built around diplomatic archives.

RENAISSANCE DIPLOMACY: COMPROMISE AS MODUS VIVENDI

Renaissance diplomacy developed among numerous small and five major Italian city-states. The Papal States (territories under the direct sovereign rule of the pope), with Rome as their capital, were located in central Italy, while the Kingdom of Naples occupied the southern part. The north was dominated by city-states with strong manufacturing and trading industries, including the Republic of Venice, the Duchy of Milan, and the Republic of Florence.

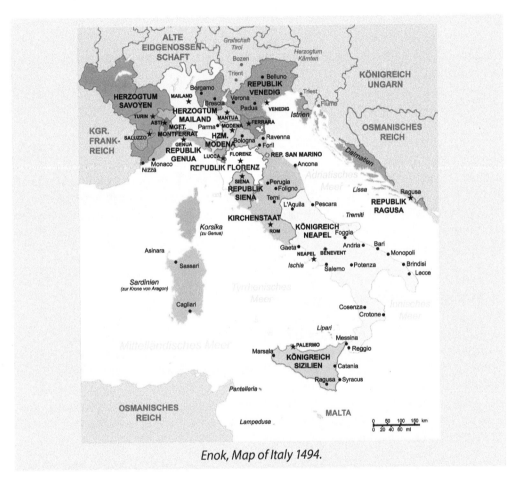

Enok, Map of Italy 1494.

During this period, the power of the Roman Catholic Church began to decline gradually; France, Spain, and the Austro-Hungarian Empire were not yet well established; and the Italian city-states developed their own way of handling internal relations. The period of Italian Renaissance diplomacy lasted between 1350 and 1494, when France invaded Italy and gradually came under strong foreign influence, most notably the Habsburgs.

The golden age of Italian Renaissance diplomacy lasted from 1454 to 1494. In 1454, the Peace of Lodi between Milan, Naples, and Florence was signed, which put an end to the wars between Milan and Venice. This period marked the first long, peaceful period after a century of war. Peace lasted until 1494, when France invaded Italy. The Peace of Lodi codified the diplomatic system among Italian city-states. In the 16th century, his type of diplomatic practice spread throughout Europe, as far as England and Spain, initially through representatives of Italian city-states in these countries and later through the exchange of ambassadors.

Two geopolitical elements created particularly fertile ground for the use of diplomacy among Italian city-states:

- the absence of a hegemonic power; and

- a strong interest in cooperating and solving problems through peaceful means.

Italian city-states were too weak to impose themselves on their neighbours. Their armed forces consisted of mercenaries who were mainly interested in earning money and surviving. The city-states could not rely on military power. This 'weakness' created an ideal space for diplomacy. The primary political tools were diplomatic combinations (Italian: *combinazioni*), which have survived until our time.

The emergence of Renaissance diplomacy

It is widely accepted in diplomatic history that the first permanent diplomatic mission was established in 1450 (representing the Duke of Milan to Cosimo de' Medici of Florence). The first envoy was Nicodemo di Pontremoli, known as 'sweet Nicodemus' in Genoa. Italian Renaissance diplomacy was commercially driven, and Italian diplomats were often bankers and traders, but they also included well-known names such as Dante, Petrarch, and Boccaccio in the 14th century and Niccolò Machiavelli and Francesco Guicciardini in the early 16th century.

The main task of resident ambassadors was to gather information and develop relations. In a world without newspapers, they became crucial intelligence gatherers. They reported on the arrival of cargoes, the situation at court, the state of an alliance, military preparations, the atmosphere in the market, and political gossip. They needed to have good manners and oratory skills. At the end of the 15th century, Ludovico Sforza, the Duke of Milan, stated that 'the worth of a prince was seen in the men he sent to represent him abroad'.[53,54]

Back in the capital, a new ministry of foreign affairs started emerging, first as an office that dealt with archives of diplomatic reports and, later on, as a more sophisticated system for collecting and analysing information, and coordinating diplomatic actions. Diplomatic reporting was the key tool for communication between diplomatic missions and the capital. Ambassadors were busy writing reports. Some of them dispatched one report each day. Many reports contained gossip about prominent personalities and life in the cities where the ambassadors served.

Venice was the most advanced state in developing reporting techniques. Besides daily reports, ambassadors had to prepare special reports called *relazioni*, which provided a strategic overview of the relationship between Venice and the country where the ambassador served. At the end of the mission, on return to Venice, each envoy was supposed to deliver a speech with detailed information about the situation in the state where they had been, and after the session, the grand chancellor would include it in the secret archive of diplomatic documents. A Venetian official explained that the reason for archiving these documents was that this way, 'documents will be saved forever, and reading them could be useful to enlighten our present rulers and those who will come into position in the future.[55] In essence, Renaissance reports were not that different from diplomatic cables released by WikiLeaks a decade ago.

New types of diplomatic practice

Diplomatic immunity: Renaissance diplomats developed the first forms of diplomatic privileges and immunities. Diplomatic immunity protects a diplomat's person, property, and communications.

Protocol: Renaissance diplomacy also inherited aspects of elaborate Byzantine ceremonies. Every detail of diplomatic protocol was negotiated as illustrated by Harold Nicolson: 'At what exact stage in the proceedings should the ambassador remove or replace his hat?'[56]

---

[53] Berridge, G. R. (2000). Francesco Guicciardini. In G. R. Berridge (Ed.),*Guicciardini's Ricordi: The counsels and reflections of Francesco Guicciardini* (p. 85).
[54] Sowerby, T. (n.d.). *The role of the ambassador and the use of ciphers*. State Papers Online 1509–1714,
[55] de Vivo, F. (2011) How to Read Venetian Relazioni, *Renaissance and Reformation* 34(1):25-59
[56] Nicolson, H. (1962). *The evolution of diplomacy* (p. 62). Collier Books.

Public diplomacy: The Italians were well aware that it was important to influence the public. In this period, especially with the invention of the printing press, we had an early form of public diplomacy. The public diplomacy of this period also relied on personal communication and prominent personalities to shape the opinions of Italian city-states.

Science diplomacy: During this period of slow and undeveloped transportation and communications, diplomats were among the few with the privilege of travelling to remote places and bringing back information. In some cases, they were involved in early 'science diplomacy', by spreading knowledge and information, especially involving countries that were expanding their colonial exploration, such as Spain and Portugal. For example, the French ambassador to Portugal, Jean Nicot, sent lemon and banana trees, as well as indigo imported from Asia, back to Paris. Nicot was also the first to bring tobacco to Paris, and it was after him that nicotine got its name.

Marriage diplomacy: Marriage negotiations among royal families and rulers were central to the diplomacy of the 16th century. One famous marriage that affected European history then was between Henry VIII, king of England, and Queen Anne Boleyn, which resulted in the creation of the Anglican Church, independent of the Vatican. Hans Holbein's painting *The Ambassadors* refers indirectly to this event.

Hans Holbein's painting *The Ambassadors*

Painted in London in 1533, *The Ambassadors* is a life-size double portrait that depicts Jean de Dinteville, once the French ambassador to the court of Henry VIII of England, with fellow diplomat Georges de Selve, bishop of Lavaur. Francis I, the king of France, had sent two ambassadors to persuade the king of England not to divorce his first wife, Catherine of Aragon. Henry VIII was dissatisfied that his marriage to Catherine had produced no surviving sons. They, like others, were unsuccessful, and history took its course, with divorce, marriage to Anne Boleyn, and the creation of the Anglican Church after the renunciation of papal authority when Henry VIII failed to secure an annulment of his marriage to Catherine.

Apart from its historical context, *The Ambassadors* explains the spirit of the time and the role of diplomats. Between the two ambassadors is a display with two shelves of objects with strong symbolic meaning. The lower shelf has earthly symbols, including a globe, a merchant's calculus book, a lute with a broken string, and a Lutheran hymn book. These items represent earthly interests and the disorderly disputes that accompany them. The two ambassadors should overcome these earthly conflicts and elevate society to the upper shelf, symbolising a stable heavenly order, represented by the tools of the science of astronomy, evoking the optimism of the Renaissance era.

RENAISSANCE DIPLOMACY: COMPROMISE AS MODUS VIVENDI

Here, we have elements of both the Renaissance and science. The function of diplomats is to act as a bridge between these two 'shelves' – the earthly and the heavenly. Although they relied on science and the power of human creativity, the presence of a skull in the painting (a distorted skull, placed in the bottom centre of the composition, visible only from a certain angle) is a reminder that pride in human knowledge and the power it gives can be perilously vain.

Hans Holbein the Younger, The Ambassadors. 1533.

Papal diplomacy

One of the main diplomatic developments in the Middle Ages was 'papal diplomacy'. The Vatican's main objective was to keep doctrinal control over Europe and suppress any actions aimed at challenging the role of the Roman Catholic Church. Papal diplomacy used various diplomatic tools, such as negotiations, treaty-making, alliances, and arbitration, and developed considerable expertise in espionage, subversion, and conspiracy.

Cristofano dell'Altissimo. Portrait of Pope Alexander VI. Vasari Corridor.

In addition to its theological and doctrinal interests, the Roman Catholic Church had complete control over the 'information technology' of the day, until the invention of Gutenberg's printing press.

One of the main reasons for this was its choice of technology for exchanging information. The decision by the Church to adopt parchment over papyrus favoured the spread of the papal-monastic network throughout Western Europe for three main reasons.

1. Unlike papyrus, which was grown almost exclusively in the Nile Delta region in Egypt, parchment was ideally suited to the decentralised agrarian-rural monastic network, because individual monasteries could remain self-sufficient, manufacturing parchment from the skins of their own livestock.

2. The collapse of the Roman Empire and its trading system resulted in the near-total disappearance of papyrus from Western Europe. Parchment thus became the dominant medium of communication, and, however inadvertently, the Roman Catholic monastic order became the chief supplier of parchment.

3. Most importantly, very few people were literate during the Middle Ages. The norm for Western Europeans, for whom much of life was violent and chaotic, was the spoken word, further reinforcing the Church's monopoly on the written word.

The relationship between parchment and the power of the Roman Catholic Church is a clear illustration of the mode of communication favouring the interests of the Church. Indeed, the clergy became the sole custodians and suppliers of written information, which had a significant impact on their share of power. The Church's monopoly on language and the written word gave it an advantage in the diplomatic scene of the Middle Ages.

The missions of other players, such as the Frankish State, represented by Charlemagne, had to have at least one clergyman because they were the only literate individuals of that age. This made it possible for the Church to be completely informed and to strongly influence the diplomatic developments of the Middle Ages.

Tools of Renaissance diplomacy

Combinazioni (diplomatic combinations) were the key modus operandi of Italian city-states. Urban centres emerged, populated by merchant and trade classes able to defend themselves. Money replaced land as the medium of exchange. Italian politics was a tangled web of alliances, conspiracies, and deceptions. The *combinazioni* was a result of this environment. As mentioned, since the Italian city-states couldn't promote their interests through military tools, they had to rely on diplomacy, and their key tool was *combinazioni*, i.e. different combinations and arrangements of players in city-states.

Other tools used included frequently changing alliances between city-states, bribery (borrowed from Byzantine diplomacy), and spying, all trademarks of Renaissance diplomacy. King Louis XI of France is said to have told his ambassadors: 'Foreign envoys, they are lying to you! Lie to them more!' On these moral deficiencies of Renaissance diplomacy, Harold Nicolson wrote: 'They bribed courtiers; they stimulated and financed rebellions; they encouraged opposition parties; they intervened in the most subversive ways in the internal affairs of the countries to which they were accredited; *they lied, they spied, they stole.*'[57,58]

The spirit of that time is captured in the famous quote on diplomacy, which you can find today in many tourist shops. Sir Henry Wotton, the envoy of the English king to Venice, said: 'The ambassador is an honest man sent to lie abroad for the good of his country', whereby 'lie' meant both 'lying abroad' (residing abroad) and 'lying' (not telling the truth).[59]

Studio of Michiel Jansz van Mierevelt. Sir Henry Wotton (1568-1639). 1620. Sotheby's.

---

[57] Nicolson, H. (1950). *Diplomacy* (pp. 43-44). Oxford University Press.
[58] Johnston, D. M. (2008). *The historical foundations of World Order: The tower and the arena* (p. 313). Martinus Nijhoff Publishers.
[59] Nicolson, H. (1950). *Diplomacy* (p. 44). Oxford University Press.

The problem was that Wotton's pun was lost in translation from English to Latin. The translation removed the ambiguity, using only the meaning 'to deceive'. This famous quotation almost ended Wotton's career. The Catholic author Scioppius paraphrased him in Latin to highlight how cunning protestant diplomats were, with Sir Wotton standing in for James I, King of England. Sir Wotton managed to save his career and remained in the service of James I after this incident. However, Sir Wotton's quotation explains a common (mis)perception of diplomacy that remains relevant in our own times. It also shows the power of double meanings in diplomacy. Solid diplomacy requires trust and correctness. If something is said, it should be the truth. Otherwise, it should not be said.

The reformation and the invention of the printing press

The Renaissance was marked by important technical advancements. The most important one (as it underlies progress in so many other fields) was the invention of the printing press in 1450 by German publisher Johannes Gutenberg.

The invention spread like wildfire, reaching Italy by 1467, Hungary and Poland in the 1470s, and Scandinavia by 1483. By 1500, around 6 million books had been printed in Europe. Without the printing press, it is impossible to imagine that the Reformation would have occurred at all.

Printmaker, Nicolas de Larmessin (1632-1694). Johannes Gutenberg.

The invention of the printing press had a considerable impact on all functions of society, including diplomacy. The Church's dominance through parchment-based writing was challenged. The Church's participation in diplomacy gradually decreased, and clergymen no longer held a monopoly on literacy. No longer were they an indispensable part of every diplomatic mission. As the American writer Mark Twain (1835–1910) once said:

> What the world is today, good and bad, it owes to Gutenberg. Everything can be traced to this source, but we are bound to bring him homage, … for the bad that his colossal invention has brought about is overshadowed a thousand times by the good with which mankind has been favored.[60]

Meanwhile…

… in the Americas:

The Aztec Empire (also known as the Triple Alliance) was a shifting and fragile alliance of three principal city-states. The largest and most powerful of the three was Tenochtitlán. The empire exerted tremendous power over central Mexico for only 100 years (from the 1420s to 1521) before falling to the Spanish conquistadors. The Aztecs were famous for the cruelties they used to remain in power, but they also knew how to 'negotiate' with neighbouring cities.

Instead of bloody wars (resulting in many captives who would be sacrificed to the gods), rivals were offered protection, stability, and economic integration into a flourishing trading system. If it surrendered, the city-state would keep its ruling dynasty and order. In return, it had to pay an annual tribute, send its soldiers to fight along with the Aztecs, and acknowledge the supremacy of the Aztec gods. If a city or region failed to surrender after peaceful diplomatic persuasion led by Aztec ambassadors, the principal city would be sacked, and its king, nobility, and warriors would be sacrificed to the Aztec gods.[61]

All kings and diplomats of subdued cities had to witness these gruesome acts. The choice laid before the enemy was quite clear, and the Aztecs relied on their reputation for ruthlessness to raise the stakes of every potential war. The empire was, in fact, a very broad alliance, kept together by fear of bloody reprisals against mutineers. However, this turned out to be the key to their demise: except for trade, nothing but fear held the whole structure together.

---

[60] Childress, D. (2008). *Johannes Gutenberg and the printing press* (p. 122). Twenty-First Century Books.

[61] World History Encyclopedia. (n.d.). Aztec warfare. https://www.worldhistory.org/Aztec_Warfare/

RENAISSANCE DIPLOMACY: COMPROMISE AS MODUS VIVENDI

At its height, the Aztec Empire consisted of nearly 400 allied cities and 3 million people. In the end, it wasn't the war that wiped out the Aztecs, but the diseases that Europeans brought with them, i.e. smallpox and measles. They lacked immunity, and the population of Tenochtitlan fell by 40% in just one year. This made them an easy target for European invaders. Mexico City was built on the ruins of Tenochtitlán, and quickly became the most important European centre in the New World.

From the Renaissance period, we move into what is widely considered to be the modern period, in which technological advancements grow apace. In the next chapter, we look at a technological innovation that transformed diplomacy and the role of the diplomat: the telegraph.

6. Telegraph diplomacy: The 'end' of distance

In the chronological sequence, the period between the end of Renaissance diplomacy (early 16th century) and the start of the golden age of diplomacy and technology (early 18th century), was shaped by the Reformation and religious wars. Central Europe came out divided, while around it new and more centralised states appeared, such as France, England, and Austria. In this intermediate period, we focus on two remarkable personalities who made a long-lasting impact on diplomacy – Hugo Grotius and Cardinal Richelieu – and one important event: the Peace of Westphalia.

By the 16th century, the Christian Commonwealth had disappeared. Without the key roles of the Pope and the Holy Roman Emperor as the ultimate arbiters in international conflicts, there were conceptual gaps in international affairs. Dutch philosopher Hugo Grotius was the first to offer a comprehensive concept to overcome these gaps. Namely, he introduced schools of natural law and international law. According to Grotius, we – as individuals – have natural rights to live and realise our potential.[62] We are entitled to these natural rights as human beings by the fact of being born. Through natural law, Grotius tried to establish a minimum moral consensus to help society build itself and overcome the divisions of the escalating religious conflicts. He started developing this approach with the idea that individuals, empowered by natural rights, are sovereign, and sovereign people create sovereign nations. This contradicted the existing idea that sovereignty was a divine right given to kings.

On this foundation, he developed the first theory of international law in his work *On the Law of War and Peace (De Jure Belli ac Pacis)* in 1625.[63] Another book by Grotius also had a lasting impact in the field of international law, namely *The Free Seas (Mare Liberum)*, where he formulated the new principle that the sea was an international territory and that all nations were free to use it for seafaring trade.[64] The teachings of Grotius influenced the construction of the Peace of Westphalia. Grotius's notion of the sovereignty of individuals has recently re-emerged in discussions around the sovereignty of personal data.

The legacy of French statesman Cardinal Richelieu also remains important today. He developed the idea of diplomacy as a continuous process and a permanent activity. The idea of the permanence of diplomacy started developing during the Renaissance

---

[62] Grotius, H. (2018). *On the Law of War and Peace*. Jazzybee Verlag.
[63] Ibid.
[64] Grotius, H. (2004). *The Free Sea*. R. Hakluyt & W. Welwood (Eds.). Liberty Fund.

with the introduction of permanent ambassadors and took full form in the time of Richelieu. Institutionally, in 1626, Richelieu established the first modern ministry of foreign affairs in more or less the same format we know today.

Richelieu also argued that diplomacy should be governed by *raison d'état* (national interest). According to this, the state's interests are primary and eternal: They are above sentiments, prejudices, and ideologies. Richelieu was also the main architect of the expansion of French diplomacy. By the 18th century, the French language had become the *lingua franca* of diplomacy and has remained so until recently.

Peace of Westphalia

Westphalia is probably one of the most frequently used historical references in modern international relations and politics. The Peace of Westphalia is the collective name for two peace treaties signed in October 1648 in the Westphalian cities of Osnabrück and Münster. These two documents ended the Thirty Years' War (1618–1648) and brought peace to the Holy Roman Empire, closing a violent period of European history during which approximately 8 million people were killed.

The Peace of Westphalia established the sovereignty of states as independent political units. The system, based on nation states, replaced the previous system, which maintained the ultimate sovereignty of the Roman Catholic Pope and the Holy Roman Empire.

The Westphalian peace negotiations began after all sides in the war were exhausted by 30 years of fighting and destruction. Ultimately, they reached a compromise that did not satisfy anyone (a good basis for compromise). The peace deal resulted from very long negotiations that lasted four years. The first six months were dedicated to agreeing on the question of precedence, which was highly controversial due to the participation of 200 rulers and more than 1,000 diplomats. Westphalia was a milestone event marking the beginning of the era of state sovereignty, which has lasted until today.

Congress of Vienna (1814/1815)

The 18th century was a period when European diplomacy tried hard to maintain a balance between the five great powers: Britain, France, Austria, Russia, and Prussia. The French Revolution and the attempts of Napoleon I to conquer Europe shook the continent's state system and paved the way for the Congress of Vienna (1814/1815).

TELEGRAPH DIPLOMACY: THE 'END' OF DISTANCE

The Congress was one of the biggest events in the history of diplomacy and laid the basis for a peace that lasted almost 100 years. It was organised in quite a relaxed atmosphere and included salons, banquets, balls, and exceptional cuisine. The Austrians paid great attention to the social aspects of the event. Two personalities especially marked the Vienna Congress: Prince de Talleyrand, the French minister of foreign affairs, and Klemens von Metternich, the Austrian minister of foreign affairs. They were the key shapers of the peace deal.

Delegates of the Congress of Vienna in a contemporary engraving by Jean Godefroy after the painting by Jean-Baptiste Isabey

The peace talks were organised to deal with Europe after the Napoleonic Wars. Talleyrand managed the impossible. Despite being on the losing side, he managed to influence the outcome of the negotiations and proved an able negotiator for the defeated France. Simultaneously, Metternich brilliantly mastered his dual roles of social representation and political leadership. His moderation in Congress produced a long-lasting European order.

The *Final Act of the Congress of Vienna* comprised all the agreements in one great instrument. It was signed on 9 June 1815. The Congress of Vienna survived the test of time – 100 years without a major global war, up until the outbreak of the First World War.

The invention of the telegraph

The 19th century was also an era of great scientific and technological breakthroughs, with the telegraph being the most important invention for diplomacy. The word 'telegraph' derives from the Greek words *têle* (at a distance) and *gráphein* (to write). Various other terms were used for the telegraph, among them 'tachygraph' from the Greek word *tachos* (speed), highlighting the temporal aspect of communication. Arguments over the name of this new device centred on its spatial vs temporal impact. Spatial impact prevailed, and we got telegraph instead of tachygraph.

The first mechanical telegraph (the semaphore) was invented in 1792 in France by Claude Chappe. It consisted of towers built in a line across the countryside. To facilitate message transmission, Chappe developed a codebook of 92 symbols. For that time, the communication speed was incredible, as the message could be transmitted from Paris to Lille (a distance of 230 km) in 10 minutes. This system was invented during the French Revolution and was used in the Napoleonic Wars (1789–1815).

By 1844, France had some 5,000 km of semaphore communication lines used mainly by the military. The only civilian use of the telegraph was for the national lottery, which, coincidentally, was also a good source of revenue for the running of the telegraph system. Great Britain and Germany also developed mechanical telegraph systems, again, mainly for military use.

We can make one historical parallel here between Chappe's mechanical telegraph and Minitel, a French predecessor of the internet, launched in 1982. Similarly, the two inventions gave France a slight advantage, but the opportunity was not fully capitalised on. Just as the mechanical telegraph was run over by the electrical one, Minitel was, much later, replaced by the internet.

The invention of the electric telegraph was more of a process than a moment of creative illumination, and different countries claim credit for the invention. The electric telegraph is often called the 'internet of the 19th century' or 'the Victorian internet'.[65] As the English historian Robert Sabine described this process, 'The electric telegraph did not, strictly speaking, have an inventor. It grew little by little towards perfection, with each inventor adding his bit.'[66] Some such additions are listed below.

- The functional principle of transmitting a message over distance was introduced by the semaphore, i.e. Chappe's 'mechanical' telegraph in 1792.

---

[65] Standage, T. (1999). *The Victorian internet: The remarkable story of the telegraph and the nineteenth century's on-line pioneers*. Berkley Books.

[66] Sabine, R. (1867). *The electric telegraph* (p. 40). Virtue.

TELEGRAPH DIPLOMACY: THE 'END' OF DISTANCE

Chappe's telegraph

- Electric batteries, invented by the Italian Volta in 1800, were an important pre-invention. The German physicist Sömmerring experimented with electro-chemical reactions and some proto-versions of the telegraph.

- In 1820, Ampère conceptualised a needle-telegraph device.

- An important step in the process of the invention of the electric telegraph was the work of the Russian diplomat Baron Pavel Schilling. During his posting in Germany, Schilling developed an electric telegraph in 1832. His invention was successfully tested in St Petersburg, where, via the electric telegraph, he connected several buildings of the Russian Chief Admiralty.

- In Britain, Cooke and Wheatstone were the first to use the telegraph for commercial purposes in 1838 by establishing a more efficient communication among the growing network of railway lines. As all the trains used the same railway lines, the exact location of each train was of the utmost importance for the normal functioning of the system. Telegraphy then spread quickly to other commercial activities, including the support of trade and financial transactions.

- In 1844, Samuel Morse (who is often incorrectly credited as the inventor of the telegraph) settled the first telegraph line between Washington and Baltimore.

Morse's main contribution to telegraphy was the invention of a special code, named the Morse Code after him, for exchanging text messages via telegraph lines.

- From 1858 to 1866, there were various attempts to lay a transatlantic cable which resulted in establishing a fully reliable link in 1866. After this, the New York and London stock exchanges were linked and frequently exchanged information. This boosted economic activities enormously.

Changes in diplomatic activities

The telegraph affected the distribution of power. Before the introduction of the telegraph, the big banking family Rothschild developed a communication system based on couriers and carrier pigeons that connected the main European economic centres. This gave them a competitive advantage, which disappeared with the introduction of the telegraph. James de Rothschild once said: 'It is a crying shame that the telegraph has been established.'[67] One of the unforeseen consequences of the telegraph was the impact it had on the emancipation of women, as they were often employed as telegraph operators. With secure jobs, greater rights, and the possibility of education, women in industrialised countries started getting more important positions.

The telegraph can be associated with concepts like techno-optimism and techno-scepticism. The former British ambassador Edward Thornton wrote:

'Steam was the first olive branch offered to us by science. Then came a still more effective olive branch - this wonderful electric telegraph, which enables every man who happens to be within reach of a wire, to communicate instantaneously with his fellow man all over the world.'[68]

This displays an optimistic view of technology. On the other hand, some rulers were cautious about the potential social impact the telegraph might have. For example, the Russian Tsar Nicolas I considered the telegraph to be 'subversive'. Afraid of its potential to disseminate information, he declined an offer by Morse to develop the country's first telegraph lines. As a result, Russia lagged greatly behind other major powers.

The control of telegraph cables thus became of crucial geostrategic importance. The control of the telegraph meant the control of information, an important element of power. Additionally, great powers laid cables to create their own global telecommunications

---

[67] Ferguson, N. (2000). *The House of Rothschild: The world's banker 1849-1998* (p. 65). Penguin.
[68] Mosco, V. (2005). *The digital sublime: Myth, power, and cyberspace* (p. 119). MIT Press.

infrastructures. Until the 1880s, the 'cable rush' was mainly inspired by commercial and telecommunications needs.

Towards the end of the century, Great Britain controlled most of the global telegraph network. Several factors led to Britain's dominant position:

- It started laying submarine cables to facilitate communication with its remote colonies, mainly India.

- Its control of the seas helped Britain to lay submarine telegraph cables without any major obstacles.

- The high cost of developing and maintaining its telegraph network was partly compensated by commercial traffic since the main strategic telegraph outposts coincided with the main trade routes (Gibraltar, Malta, Cyprus, Alexandria, Aden, Singapore, Hong Kong, etc.).

Britain established a starting advantage that was difficult for many countries to catch up with in the following decades. Other countries realised both the importance of having a telegraph network and the extent of British dominance relatively late. Although France pioneered the development of the telegraph, it was a latecomer in developing a global telegraph cable network. It was only after a series of crises (Tonkin, Siam, and Fashoda) that the French started taking seriously the lack of a telegraph network and British dominance in this area.

During the 1898 Fashoda Incident in Africa, French plans to control Africa from the west (Dakar) to the east (Djibouti) clashed with British ambitions to establish a north–south control over the continent, from Cairo to Cape Town. These two colonial projects collided in Fashoda, a small village in present-day Sudan. Even though France had an advantage on the field, it didn't have the means to communicate this to its headquarters. On the other hand, the British conveyed false information to London about the difficult position of the French troops in Fashoda. The technological advantage was so strong that the French officials were forced to ask their British counterparts to send a message to Paris via the British telegraph.[69] After this crisis, France and Germany realised that they had to close the 'cable gap' and started developing their own global cable networks.

The United States also started to emerge as a global political and economic power, which could be seen in the extent of its global cable network. Unsuccessful attempts to establish a transatlantic link are often cited as one of the reasons that the USA purchased Alaska, then known as Russian America. Faced with difficulties in establishing a

---

[69] Bates, B. (1984). *The Fashoda Incident of 1898: Encounter on the Nile*. Oxford University Press.

transatlantic link, Western Union president Hiram Sibley urged the purchase of Alaska to establish a 16,000-mile land-based wire between the USA and Europe. This terrestrial telegraph scheme was abandoned in 1868 when the transatlantic cable proved successful. Alaska, though, remained part of the United States. The telegraph gradually became part of international negotiations and diplomatic tactics.

A new topic on the diplomatic agendas

The telegraph appeared early on diplomatic agendas. In the mid-19th century, the first international agreements were signed to manage telegraph communication.[70] The most developed network of bilateral agreements took place in Germany, which at the time was divided into many small states. In 1849, Prussia and Saxony concluded the first bilateral agreement, and others followed. Only one year later, the German states and Austria established the Austro-German Telegraph Union (AGTU). Other European countries started concluding bilateral agreements, which led to the establishment of the West European Telegraph Union (WETU) in 1855, a regional organisation that included Belgium, France, the Kingdom of Sardinia, and Switzerland. Other countries subsequently joined WETU.

In the end, bilateral arrangements could not keep pace with the intensity of technological developments. The need for a comprehensive multilateral arrangement was obvious, and the first multilateral arrangement was adopted in 1865 in Paris with the establishment of the International Telegraph Union (ITU). In 1868, the International Bureau of Telegraph Administration was established in Bern; it is considered to be the first permanent international organisation.

One of the main international legal issues raised in 1865 at the International Telegraph Conference held in Paris was the neutral status of submarine telegraph cables. France, later joined by Germany and other states, requested the protection of submarine telegraph cables in case of war. The main opponent was Great Britain, since it both controlled most submarine cables and the technology for managing them (including cable-cutting tools). Britain combined collecting information for the laying of submarine telegraph cables with many naval and scientific expeditions, such as that of the HMS *Challenger* in 1872–1876, which shows how developments in scientific, naval, and communication technologies overlapped and could catalyse each other. In the end, cables remained outside the regulations covering the conduct of war.

Many important political decisions that influenced the future development of the telegraph were adopted at the next diplomatic conference, the International Telegraph

---

[70] ITU. (n.d.). The earliest international telegraph agreements. *ITU Portal.* http://handle.itu.int/11.1004/020.2000/s.139

Convention of St Petersburg, in 1875. One of the most controversial issues was controlling the content of telegraph communication. While conference participants from the USA and Great Britain promoted the principle of the privacy of telegraph correspondence, Russia and Germany insisted on limiting this privacy to protect state security, public order, and public morality. A compromise was reached through an age-old diplomatic technique: diplomatic ambiguity.

While Article 2 of the St Petersburg Convention guaranteed the privacy of telegraph communication, Article 7 limited this privacy and introduced the possibility of state censorship. The United States, which did not participate, refused to sign the convention because of 'the censorship article'. Introducing new topics on diplomatic agendas influenced the organisation of diplomatic services and led to the emergence of new posts in diplomatic missions, such as military attachés, as well as diplomats in charge of economic and cultural affairs.

The use of new tools in diplomacy

It was in the 1850s that the telegraph started to be used as a tool in diplomatic services. The telegraph was initially used in diplomacy for internal communication between diplomatic missions and headquarters. This also included communication regarding personal, ceremonial, and organisational matters. During the Congress of Paris (1856), British representatives received instructions from Prime Minister Palmerston through coded telegrams.

In 1866, the US State Department sent a cable to the US Mission in Paris. The encrypted message that passed through British and French cables did not prove to have any diplomatic importance, but it did go down in the history of diplomacy and technology as one of the most expensive tech experiments in diplomacy. The cost of dispatching this telegram was US$20,000, while the total annual budget of the US State Department at the time was US$150,000. This huge bill triggered a court case between the State Department and the telegraph company. Eventually, after the Supreme Court's decision, the US government was forced to pay for this expensive diplomatic experiment.[71]

By the end of the 19th century, the telegraph was being used in day-to-day diplomatic activities, which was met with mixed reactions. The British ambassador to Venice, Sir Horace Rumbold, commented on the negative impact of the telegraph on the independence of diplomats in their function: 'the telegraphic demoralisation of those who

---

[71] Weber, R. E. (1993). Seward's other folly: America's first encrypted cable. *Cryptologic Quarterly* 12(2), 81-103.

formerly had to act for themselves and are now content to be at the end of the wire.'[72] Diplomats also frequently complained about the lack of time for a proper and analytical approach to diplomatic activities.

The telegraph also became part of diplomatic tactics. Three telegrams had a considerable impact on the development of international relations and, to a certain extent, changed the course of history:

1. The Ems Telegram is usually considered an example of how technology could be used to achieve strategic military objectives. In the 1870s, Bismarck's objective was to unify Germany, and his first step was to start a war with France. How did a dispatch help Bismarck achieve his aim? In a time of crisis in the relations between France and Prussia, the Prussian king, who was not very keen on starting a war with France, sent Bismarck a telegram from Ems in which he reported on his meeting with the French ambassador and asked Bismarck to inform the diplomatic corps and press. BBismarck, who was looking for the cause to start war with France, rephrased the king's conciliatory message into public provocation which triggered the war and led to the German unification of 1871.

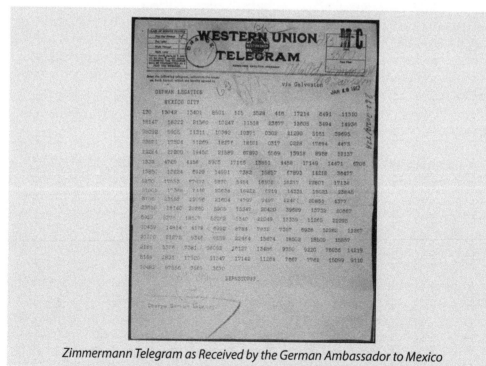

Zimmermann Telegram as Received by the German Ambassador to Mexico

---

[72] Nickles, D. P. (2003). *Under the wire: How the telegraph changed diplomacy* (p. 45). Harvard University Press.

2. The Zimmerman Telegram became part of history because it influenced the USA to abandon its neutrality and enter the First World War in 1916. The telegram was sent by the German foreign minister to the German ambassador in Mexico, allegedly asking him to offer Mexico parts of California and Texas in exchange for Mexico, thereby joining Germany in the war. The telegram was intercepted by British intelligence and rewritten in such a way as to provoke a strong anti-neutrality sentiment in the USA. This ultimate goal was achieved, and the USA entered the war.

3. The July Crisis telegram was sent during the July Crisis of 1914 just before the start of the First World War. The major problem, which led to the failure of diplomacy, was the diplomats' inability to cope with the volume and speed of electronic communication. As Prof. Stephen Kern noted: 'This telegraphic exchange at the highest level dramatised the spectacular failure of diplomacy, to which telegraphy contributed with crossed messages, delays, sudden surprises, and unpredictable timing.'[73] Diplomats 'failed to understand the full impact of instantaneous communication without the ameliorating effect of delay'.[74]

In the example of the July Crisis, the important rationale is that speed and immediacy do not necessarily provide positive results. On the contrary, they can likely lead to reckless moves and miscommunication.

The most important aspects of the telegraph

1. The need for urgent replies
The speed of message transfer has been linked to urgency. If a message travels for months, one can afford time to prepare a proper response. This situation changed with the telegraph. The immediacy of sending back messages required immediate responses from diplomats abroad. This led to potentially hasty and, sometimes, ill-prepared responses.

2. The problem of coordinating communication
Urgency led to the problem of coordinating communication. Very often, telegrams would arrive in the wrong order, creating considerable confusion with serious consequences, as was the case before the First World War. During a delicate exchange on the Alabama dispute, US foreign secretary Granville warned British prime minister

---

[73] Kern, S. (1983). *The culture of time and space*, 1880-1918 (p. 268). Harvard University Press.
[74] Ibid. (p. 276)

Gladstone of this risk: 'This telegraphing work is despairing. It will be a mercy if we do not get into some confusion.'[75]

3. The need to prepare concise messages

The telegraph was an expensive medium, so the content of each telegram had to be carefully considered. Diplomats had to improve the quality of diplomatic reportage, abandon long and descriptive memos, and master the skill of concise and precise writing. The telegraph is a good example of how technology influenced style and etiquette.

4. The emergence of foreign policy bureaucracy

Although the first ministries of foreign affairs were established earlier, their number increased at the end of the 19th century due to bureaucratic expansion. French diplomacy grew from 70 diplomats in 1814 to 170 diplomats a century later. The Habsburg Empire had 51 diplomats in the mid-19th century and 146 in 1918. Foreign ministries achieved the shape that they, more or less, have retained until today. They mainly consisted of geographical departments handling bilateral relations with various countries. Entrance exams gradually replaced family ties as the main recruitment method. New diplomats were trained in diplomatic academies and a sense of professionalism started to prevail.

5. The centralisation of diplomacy

Easier communication via the telegraph and well-equipped ministries led to the centralisation of diplomatic services. Previously, independent diplomatic missions came under the control of their headquarters, and instructions could easily be sent via the telegraph.

1814–1914: Statesmen (from leaders to followers)

Statesmen and diplomats were essentially unprepared for the sudden appearance of this new technology, which overwhelmed them and contributed to the fact that they ceased to be the creators of policies and became followers of events that often seemed to be getting out of their control. Unprepared to handle the new technology, diplomats who used to meet and manage international peace through direct communication suddenly became involved in the frenetic world of modern diplomacy conducted via telegraphs and telephones.

The period after the Vienna Congress was decisive for diplomatic developments and was heavily influenced by technology (the telegraph and telephone). In this period, the basis was laid for the diplomacy we have today: embassies, ministries, communication,

---

[75] Ramm, R. (Ed.). (1998). *The Gladstone-Granville correspondence* (p. 240). Cambridge University Press.

TELEGRAPH DIPLOMACY: THE 'END' OF DISTANCE

and diplomatic reporting. By studying this period, we can learn how not to repeat the same mistakes and how to create more optimal solutions for potential confusion and conflicts.

Meanwhile…

…in China:

The first telegraph lines in China were established during the 1860s, when European imperial powers used them to connect with their colonies. The government of the Qing dynasty was initially very reluctant to establish their own telegraph, fearing that the European powers might use them. The first Chinese line was established in 1871 from Hong Kong to Shanghai by Denmark's Great Northern Telegraph Company.

This also meant introducing the telegraphic code for Chinese, which greatly differed from the alphabet code. A few years later, the Chinese built a line mainly used for military purposes. In 1876, China's first telegraph school was opened. Its students learned not only how to operate, but also how to construct telegraphs and lay their lines. In this way, China secured an educated class of telegraph operators.

In 1881, the government established the Imperial Chinese Telegraph Administration (ICTA), which took control over all existing networks in the country, except foreign ones. By 1900, it had 14,000 miles of telegraph lines connecting all the cities along the coast. However, its prices were high and transmission delays were frequent, so it did not have as deep an impact on Chinese society as it did in many other countries.

The invention and proliferation of the telegraph had a significant impact on diplomacy, and the extension of telegraph infrastructure would continue in the latter part of the 19th century. In our next chapter, we look at another technology that would speed up long-distance communication considerably: the telephone.

7. Telephone diplomacy: Dialling the 'red line'

The telephone, radio, and telegraph constitute the three most important inventions that have shaped communication up until today. The telegraph delinked communication from physical transportation and travelling, the telephone transferred voice over great distances, and the radio delinked communication from almost any physical medium.

During the late 19th century, the process of accelerated technological advancement and imperial expansion (and, for some, decline) of the preceding decades continued. The British and Russian empires expanded considerably in this period, while those such as the Ottoman, Spanish, and Mughal empires continued to slowly decline. The period saw vast emancipation, with the abolition of formal slavery across vast areas of the world, as well as the abolition of serfdom.

Enabled by growing rail and communication networks, the period also saw intensified permanent settlement, mostly by Europeans, such as westward within the United States and in Australia in the British Empire. Tens of millions of people migrated and settled around the world in this period, mostly from Europe to the Americas.

The middle of the 19th century saw the period of modernisation, known as Tanzimat, in the Ottoman Empire, while European powers consolidated colonial power and imperial rule over almost all of Africa, and the French established the rule of the Nguyễn dynasty of Vietnam, incorporating it in 1887, along with Cambodia, into French Indochina. The defeat in the Crimean War in 1856 did not stop the Russian Empire from accelerating its expansion into Central and Northeast Asia.

These huge imperial expansions coincided with and perhaps provided some impetus for the technological and industrial advancements of the period. The consolidation of steam power in industrial processes, known as the Second Industrial Revolution, the spread of ironworks, and the beginning of the widespread use of electricity, all catalysed industrial developments further, as vast resources came under the control of capitalist empires and republics. As machines of imperial and economic order grew, so did the demand for better, and faster, communications technologies.

These inventions have also strongly influenced diplomacy. The telephone enabled close contact between heads of state, including via various 'red lines'. The radio had a strong impact on communications geopolitics. Some diplomatic issues – security, privacy, and neutrality – that were raised in discussions surrounding telephone and wireless communications are still being discussed today in the context of digital policy.

Invention of the telephone

As with many innovations, the idea for the telephone came much earlier than the invention saw the light of day. The general and widely accepted view is that the inventor of the telephone was Alexander Graham Bell, but this invention was a culmination of the work of several individuals.

Antonio Meucci, (1808 - 1889), inventor of the telephone

One of the first successful experiments with telephony was conducted by the Italian immigrant Antonio Meucci back in 1854. He invented the first voice-communicating device in 1854 and called it *teletrofono*.[76] Meucci experimented with telephony 20 years before Bell, but went bankrupt and couldn't afford to patent it. He was recently officially acknowledged as the inventor of telephony by the US Congress.[77] In 1861, German Philippe Reis created a prototype of the telephone that transmitted voice sounds

---

[76] Engineering and Technology History Wiki. (n.d.). Antonio Meucci. https://ethw.org/Antonio_Meucci

[77] H.Res.269 — 107th Congress (2001-2002): Expressing the sense of the House of Representatives to honor the life and achievements of 19th Century Italian-American

electrically over distance. Elisha Gray invented the tone telegraph (harmonic telegraph) in 1875, and was granted the US patent for Electric Telegraph for Transmitting Musical Tones.

Although he was not the first to experiment with telephonic devices, what makes Bell important in the history of telephony is that he had enough capital and creativity to make telephony a practical utility. He, and the companies founded in his name, were the first to develop commercially practical telephones around which a successful business could be built. It can thus be argued that Bell invented the telephone industry.

Anti-telephone campaign

Bell encountered opposition from Western Union, a leading telegraph company at the time.[78] He wanted to sell them his patents and technology for the telephone, but they thought it was laughable:

> The idea is idiotic. Furthermore, why would any person want to use this ungainly and impractical device when he can send a messenger to the telegraph office [...] Technically, we do not see that this device will ever be capable of sending recognizable speech over a distance of several miles.[79]

When Western Union could not stop the development of telephony, they signed a contract with Bell in 1879, stating that the telephone should only be used for personal conversations, while the telegraph would remain the main communications tool for businesses. Obviously, such a clause could not be sustained, as the telephone became a frequently used communication tool by stockbrokers, bankers, lawyers, doctors, and other professionals who depended a great deal on communication.

Telecommunications network in 1900

The diffusion of this new technology was uneven and was influenced by various technical, economic, and social factors. In 1900, an early communications divide, similar to the modern digital divide, was obvious. A considerable difference between the USA and Europe was present, as was a north–south division within Europe itself, with the densest

---

inventor Antonio Meucci, and his work in the invention of the telephone. https://www.congress.gov/bill/107th-congress/house-resolution/269/titles

[78] Elon University. (n.d.).*Imagining the internet: A history and forecast, 1870s–1940s: Telephone.* Elon University. https://www.elon.edu/u/imagining-time-capsule/150-years/back-1870-1940/

[79] Ibid.

telecommunications network in Sweden and the least dense in Italy. Sometimes, such as in the case of France and Germany, which had similar levels of overall technological development, the same patterns did not apply to the telephone adaptation: Germany had three times higher telephone penetration than France.

The importance of the telephone in diplomacy

For a long time, the widespread use of the telephone was limited because it was difficult to sustain the strength of telephone signals over longer distances. It took a few decades to get the first direct telephone line between New York and San Francisco (1914), and even longer for transatlantic telephone lines between the USA and Europe (1956).

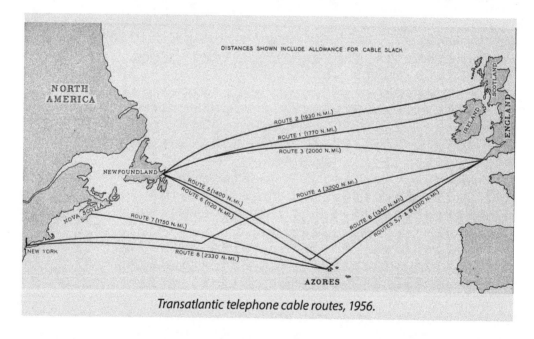

Transatlantic telephone cable routes, 1956.

The telephone had an enormous social impact. It became an integral part of private, professional, and official lives in most societies. The real impact of the telephone on international relations was felt after the Second World War. State leaders, especially those of the USA and the USSR, started using the telephone to avoid further escalation in international crises. The telephone played an important role in several international crises, including:

- The Six-Day War in the Middle East in 1967.

- The India–Pakistan crisis in December 1971.

- The Arab–Israeli war of 1973.

- The Invasion of Afghanistan by the USSR in 1979.

During the Cuban Crisis (1961), the USA and the Soviet Union were on the brink of war. The existing methods of communication between Washington and Moscow were too slow for the events taking place: It took Washington nearly 12 hours to receive and decode Nikita Khrushchev's 3,000-word initial message. By the time a reply had been written and edited by the White House, Moscow had sent another, tougher message. Under severe time pressure, both leaders ultimately communicated through the media. After the crisis was resolved, the hotline proposal became an immediate priority. The first message was sent on 30 August 1963, including numbers and an apostrophe to ensure the connection worked properly. The USA sent the message: 'The quick brown fox jumped over the lazy dog's back 1234567890.' This is a common test message, as it includes all 26 letters of the English alphabet. The first official use was to announce the assassination of John F. Kennedy to Russia.[80]

It is said that the 1958 novel *Red Alert* by Peter Bryant (which inspired Stanley Kubrick's film Dr. Strangelove) gave government officials the idea to get in touch directly, by showing the benefits of fast and direct conversations.[81] Although it was called the 'red telephone', it was actually first a telegraph, followed by a fax machine, and, lately, an advanced email system. The phrases 'red telephone' or 'hotline' describe direct communication between the leaderships of two countries in a crisis. The red telephone symbolised importance and exclusivity in relations between countries. It made other countries, mainly the UK and France, establish exclusive communication lines with Moscow. Today, leaders maintain exclusive contact via mobile phones.

The invention of radio communication

The rush for new inventions marked the whole of the 19th century. Towards the end of the century, wireless communications became the new scientific frontier. The very first use of radio-transmitted coded information was the result of the work of James Clerk Maxwell (Britain) and Heinrich Hertz (Germany) and their pioneering experiments with electromagnetic waves. Maxwell provided the theoretical basis, which Hertz confirmed through experiments. Their scientific discovery was used for subsequent inventions in the field of wireless communications, starting with wireless telegraphy, via wireless telephony, and concluding with radio broadcasting.

---

[80] Electrospaces. (2012, October 28). The Washington-Moscow hotline. *Electrospaces.* https://www.electrospaces.net/2012/10/the-washington-moscow-hot-line.html

[81] Bryant, P. (1958). *Red alert*. Ace Books.

Other inventors who worked on wireless communication were Eduard Branly, Oliver Lodge, and Alexander Popov. Additionally, Nikola Tesla was particularly notable for designing both emitters and receptors for electromagnetic waves. He used the term 'wireless telegraphy' ('radio' as a term that appeared only after the First World War). Tesla had also designed a technical solution for transmitting power wirelessly.

The Indian professor Jagadis Chandra Bose from Calcutta was a pioneer in the research of radio technology, and, for the first time ever, demonstrated wireless communication using radio waves. He said, 'It is not the inventor but the invention that matters', and he never patented his work. Bose believed knowledge should be available to everyone and not constrained by patenting.[82]

Capitalising on the previously described theoretical inventions, Italian scientist Guglielmo Marconi invented a wireless communication device. In 1897, he registered a patent for his wireless telegraph. In addition to his invention, Marconi had other advantages and talents that helped him disseminate this device. He had family ties and business links with Great Britain, the world leader in telecommunications at the end of the 19th century. His marketing and public relations talents helped him secure a few lucrative deals with the British Admiralty and British shipping companies. In 1907, Marconi's wireless telegraph system became a public service for transatlantic exchanges between Europe and the USA. The main users of Marconi's wireless telegraph were the British and Italian militaries.[83]

Wireless geostrategy

Great Britain, together with the USA, often held a monopoly on cable communications. At one point, together, they controlled over 70% of the global telegraph cable infrastructure. The countries that were lagging behind, mainly Germany and France, used radio communication as a way to bypass this monopoly. This was particularly important for Germany, whose weak cable-based communications infrastructure could not match its increasing geostrategic ambitions.

The German government supported research and development in the field of wireless communications. In 1903, two pioneering producers of wireless telegraphy sets, AEG and Siemens-Halske, merged under government pressure into a new company – Telefunken.

---

[82] Krishnan, R. (2020, November 30). J.C. Bose – 'Father of Radio Science' who was forgotten by West due to his aversion to patents. *ThePrint*. https://theprint.in/features/j-c-bose-father-of-radio-science-who-was-forgotten-by-west-due-to-his-aversion-to-patents/552556/

[83] Sherrow, V. (2004). *Guglielmo Marconi: Inventor of radio and wireless communication*. Enslow Publishers.

Guglielmo Marconi

In an attempt to take maximum advantage of his invention, Marconi established a monopoly by preventing operators who used his system from communicating with radios developed by other companies. Germany tried challenging this monopoly at two International Radiotelegraph Conferences (in 1903 and 1906). Although most countries were against Marconi's monopoly, the objections of Great Britain and Italy kept the status quo until 1912 by not forcing Marconi to make his system interoperable with other producers.

Telegraph standardisation battle

Things changed in 1912 after the British ship *Titanic* sank. One of the reasons that more passengers from the *Titanic* were not saved was that a nearby ship, the Californian, which was five miles away, did not receive the distress radio signal from the *Titanic*. The first ship that came to save the passengers was the *Carpathia*, which was 45 miles away.

Two major companies provided the equipment and operators: The Marconi Company in New York City and Telefunken in Germany. The *Titanic* was subscribed to Marconi. [The] …The Marconi Company issued an edict that any operator who "talked" to a Telefunken ship would be immediately relieved of duty upon his return. Telefunken, in turn, issued the same order to their operators. [...][84] This is why the SOS from the *Titanic* went unanswered by the *Californian*, a Telefunken ship, which we now know was only

---

[84] Ryan, P. (2012) The ITU and the Internet's Titanic Moment (July 13, 2012). Stanford Technology Law Review, Vol. 2012, No. 8, Available at SSRN: https://ssrn.com/abstract=2110509

miles away. The *Carpathia*... a Marconi ship... heard the SOS and was able to respond, even though they were some distance away.[85]

The Titanic disaster had a far-reaching impact on global telecommunications policy. The International Radiotelegraph Conference held in 1912 ended Marconi's monopoly by introducing the principle of interconnectivity among radiotelegraph systems.

The rising importance of media in diplomacy

The period between 1814 and 1914 was a golden period both for diplomacy and telecommunications. The 'Long Peace' was initiated by the Congress of Vienna (1814), which introduced the Concert of Europe as a way to deal with international crises. The Concert of Europe was a vague consensus among European monarchies favouring the preservation of the territorial and political status quo. During this period, the important topics were privacy, security, and the neutrality of telecommunications.

The new political environment, influenced by the development of communications technology, had a considerable impact on questions of war and peace. Both diplomats and the military had to adjust their methods to the changing environment.

The emergence of the radio and the rising importance of the press led to a different operational environment for diplomats.

The public began challenging the closed and exclusive club of negotiators assembled through the Concert of Europe. Mass literacy and the growing number of newspapers triggered the development of public opinion. Towards the end of the 19th century, diplomats became increasingly concerned about the reactions of their domestic audiences, who had become well-informed about diplomatic activities. The development of public opinion also put pressure on governments and diplomatic services. Following this new threat, monarchies and governments introduced censorship and started using newspapers for foreign propaganda.

The US invasion of Cuba in 1898 was the first example of the importance of the media in international relations. Many historians believe that if it had not been for the media's propaganda, this war could have been avoided. This is effectively illustrated in a story involving a journalist and the US media mogul William Randolph Hearst. The journalist sent the following message to Hearst: 'Everything is quiet. There is no trouble here.

---

[85] Dawson, K. (1999, January 3). Browser wars of the wireless telegraphy age. *TBTF*. https://tbtf.com/resource/telegraph-browser-wars.html

There will be no war. I wish to return.' The reply from Hearst was: 'Please remain. You furnish the pictures, and I'll furnish the war.'[86]

Meanwhile...

...in Africa

The development of the first telephone networks coincided with the great push of European powers to establish more solid control over inland Africa and eventually divide the whole continent into their colonies. This effort was characterised by the construction of railway lines, closely followed by telegraph and telephone lines. Already in the 1890s, there were towns in Africa with telephone service.

However, the development of these telephone lines was primarily to serve the needs of the European imperial rulers and not the needs of the local population. The idea behind establishing telephone lines was to link African colonies with the capitals of the British, French, German, and other empires. Good connections were created between regions under the control of one empire, but not with neighbouring towns or regions ruled by a different European power.

This became painfully evident when, in the early 1960s, African nations got their independence. One famous example is that the telephone connection between Brazzaville and Kinshasa – standing on opposite sides of the Congo River – needed to be routed via Paris and Brussels. This was one of the problems that stood in front of the Pan-African Telecommunications Network (PANAFTEL), formed in 1962 in Dakar, Senegal. However, it took more than a decade to implement the first steps of the project, and African nations started to become connected by a combination of copper wire and microwave links. The first submarine cables reached northern Africa in 1956, but it wasn't until 1969 that they reached the sub-Saharan part of the continent. The slow dissemination of landlines was one of the main reasons why mobile telephone technologies became the preferred option among users, and thus developed at an astonishing speed.

While the telephone revolutionised direct communication practices in the late 19th century, our next technologies changed how diplomacy could reach a wider audience, inaugurating the role of a diplomat as a public figure and marking the beginning of public diplomacy: radio and television.

---

[86] Campbell, W. J. (2001). *Yellow journalism: Puncturing the myths, defining the legacies* (p. 71). Praeger.

8. Public diplomacy: Going live on TV and radio

The processes of the previous century continued to grow apace at the beginning of the 20th century, but the period of stable and expanding imperial orders was hugely disrupted by multiple events that caused cataclysmic political changes: the First World War, the Russian Revolution, the Great Depression, the Second World War, decolonisation, and the Cold War. These events surrounded, used, and accelerated the growth of mass communication technologies.

The growth of mass media also worked to shape public interest and create consumer desires, as large and growing middle-class populations, particularly in the West, became bases for growing markets of consumer products. This is particularly visible in TV advertising, and the crossover between state propaganda and mass marketing is demonstrated by their having been pioneered by the very same group of public relations experts, such as Edward Louis Bernays.

The period also saw widespread women's suffrage, mostly following the First World War, but with some notable exceptions, with voting populations in many countries effectively doubling (some countries, such as Australia, would only initially enfranchise non-indigenous women to vote, with the right to vote for indigenous women not being established until decades later).

The invention of wireless communication

In the previous chapter, we discussed the invention of wireless communication, a technology that had many inventors, such as Maxwell, Hertz, Branly, Lodge, Popov, Tesla, Bose, and Marconi. They provided the scientific foundation on which radio communication technology was developed as a way of transferring sound over radio waves.

Transferring sound over radio waves

Both Guglielmo Marconi and Reginald Fessenden were attempting to create a wireless telephone, by combining two inventions: a telephone and wireless communication. In line with the principle of unintended consequences, their pursuit resulted in something else. Instead of a wireless telephone (a two-way communication), we got a 'radio music box', a form of one-way communication.[87]

The simplicity of this device was the very reason why radio broadcasting proved to be a powerful communication tool. Fessenden carried out the first-ever broadcast of a radio programme of speech and music. On Christmas Eve of 1906, from the radio station in Bryant Rock, Massachusetts, he transmitted a violin performance and read a Christmas greeting himself. The wireless broadcast reached ships over a radius of more than 150 miles.

The Marconi Company began broadcasting from Chelmsford, Essex, in 1920. The Marconi engineers broadcast music instead of simple, dull test messages. Due to the fear of commercialisation of the new medium, and the military's claim on unoccupied airwaves, the British Post Office placed a ban on the daily broadcasts and ruled that experimental broadcasts must be individually licensed. In 1922, the British Post Office permitted the Marconi Company to broadcast 15 minutes of musical programmes per week. Soon, another experimental station was set up in London. The British realised the potential of broadcasting and the need for its regulation. On 18 October 1922, the BBC (British Broadcasting Company Ltd.) was established, to monitor the development of the industry. By the 1940s, 83% of Americans, and almost 35% of British people owned a radio. A decade later, this number doubled, but then slowly, television started replacing radio as the favourite medium.

The power of radio

For centuries, the main medium for public diplomacy was the printed word. As we have seen, Gutenberg's invention of print played an important role in the diplomatic and religious struggles of the Middle Ages. The power of radio was discovered quite early, between the First and Second World Wars. For the first time, politicians could address the wider population directly through radio, without having their message filtered through the press. For example, Franklin D. Roosevelt introduced fireside chats – radio talks addressing the problems and successes of the Great Depression. In March 1933, CBS announced: 'The president wants to come into your home and sit at your fireside for a little fireside chat.' Roosevelt's press secretary praised the radio, saying, 'It cannot misrepresent or misquote.'[88] It was during this period that the basic radio propaganda infrastructure was established.

---

[87] Kurbalija, J. (2021). *Radio and TV broadcasting: Diplomacy going live.* https://www.diplomacy.edu/histories/radio-and-tv-broadcasting-diplomacy-going-live/

[88] Biser, B. (n.d.). The fireside chats: Roosevelt's radio talks. *The White House Historical Association.* https://www.whitehousehistory.org/the-fireside-chats-roosevelts-radio-talks; Roosevelt, F. D. (2003). *The fireside chats of Franklin Delano Roosevelt.* Project Gutenberg.

PUBLIC DIPLOMACY: GOING LIVE ON TV AND RADIO

Joseph Goebbels, the German propaganda minister, understood the art of persuasion. He wanted to use the potential of radio to broadcast Hitler's messages. He declared radio 'the eighth great power' (alluding to Napoleon's view that the press was the seventh great power). Goebbels made sure that everybody had an affordable radio receiver and distributed millions of cheap radio sets to German citizens.[89]

Winston Churchill at a BBC microphone about to broadcast to the nation on the afternoon of VE Day, 8 May 1945.

The previously established technological dynamism continued during and after the Second World War. It is said that the radio was the most significant tool that influenced the outcome of that war. Radio was used both as a propaganda and a morale-boosting medium by all sides involved in the conflict. For example, Winston Churchill spoke directly to the people via the BBC during the war, and more than half the adult population tuned in to listen while he was on air.[90] Another example of the power of radio broadcasting was Orson Welles's 1938 radio adaptation of H.G. Wells's *War of the Worlds*. His 'announcement' of the Martian invasion of the Earth reportedly created panic in the United States.

The panic, however, was actually created by the American newspapers. Radio took revenue away from print, damaging the newspaper industry. In return, they took the

---

[89] Marsh, A. (2021, March 30). Inside the Third Reich's Radio. *IEEE Spectrum*. https://spectrum.ieee.org/inside-the-third-reichs-radio; Bergmeier, H. J. P., & Lotz, R. E. (1997). *Hitler's airwaves: The inside story of Nazi radio broadcasting and propaganda swing, Volume 1*. Yale University Press.

[90] BBC. (1939, October 1). *Winston Churchill's first wartime broadcast*. https://www.bbc.com/historyofthebbc/anniversaries/october/winston-churchills-first-wartime-broadcast

opportunity to discredit radio as the news source. The newspaper industry sensationalised the panic created by Welles's episode, to prove to advertisers and regulators that radio management was irresponsible and not to be trusted.[91]

Cold War: Ideological battle via radio waves

The radio broadcasting infrastructure, established during the Second World War, found another use during the Cold War period. International broadcasting increased, containing propaganda disguised as news, with Communist and anti-Communist states attempting to influence each other's domestic populations.

The United States established the Voice of America in 1942, which continued operating after the war ended. Radio Free Europe was established in 1950, and Radio Liberty in 1951. In 1947, the Voice of America started broadcasting in the Soviet Union as a part of US foreign policy to fight Soviet propaganda. This resulted in the aggressive jamming of Voice of America's broadcasts by the Soviets.

While objecting to western radio broadcasting, the USSR developed a similar, if not even more powerful, radio system covering almost the entire globe: Radio Moscow. In Great Britain, through Churchill's mastery of the medium, the BBC also underwent a big expansion. At the end of the war, the BBC increased the number of broadcasting languages to 45, along with its transmitting power. Previously, the Axis Powers had also broadcast in many languages.

Radio broadcasting as a diplomatic issue

Radio broadcasting entered international relations and the diplomatic arena early. Many governments worldwide realised the power of radio in reaching foreign audiences, and a new way of conducting diplomacy was born: public diplomacy. In 1936, the League of Nations drafted and signed a treaty called: *The International Convention Concerning the Use of Broadcasting in the Cause of Peace*, where the states agreed to prohibit the use of broadcasting for propaganda or the spreading of false news.[92] The

---

[91] Pooley, J., & Socolow, M. J. (2013, October 28). The myth of the *War of the Worlds* Panic. *Slate*. https://slate.com/culture/2013/10/orson-welles-war-of-the-worlds-panic-myth-the-infamous-radio-broadcast-did-not-cause-a-nationwide-hysteria.html; Schwartz, A. B. (2015). *Broadcast hysteria: Orson Welles's War of the Worlds and the art of fake news*. Hill and Wang.

[92] World LII. (n.d.). The League of Nations, Treaty 4319, The International Convention Concerning the Use of Broadcasting in the Cause of Peace, LNTSer 80; 186 LNTS 301, (September 23, 1936), http://www.worldlii.org/int/other/treaties/LNTSer/1938/80.html

effect of this Convention was limited since Germany, Italy, and Japan weren't parties. In addition, China, the USA, and the Soviet Union didn't ratify it either.

Another diplomatic issue was the jamming of radio programmes, which became an important issue in diplomatic negotiations during the Cold War. Politicians on both sides had an issue with the notion that the ideas could be transmitted beyond the borders of any country.[93] The Western bloc believed that states could not exercise sovereignty over ideas broadcast by radio. On the other hand, the Eastern bloc claimed that states had a duty to jam radio signals when they transmitted ideas that criticised the socialist order. These positions mark the outlines of the debate over freedom of information. The USSR heavily used radio jamming to prevent its citizens from listening to 'politically dangerous' broadcasts from the BBC, the Voice of America, and other western broadcasters. The USA also jammed Radio Moscow, Radio Pyongyang, and Radio Havana, Cuba.

The UN resolution of 1950 condemned the Soviet Union's deliberate interference with the radio signals as 'a violation of the accepted principles of freedom of information'.[94] Nevertheless, the Soviet Union continued to jam foreign broadcasts. The practice of jamming remained a diplomatic issue until the end of the Cold War.

The future of radio

Radio broadcasting confirms that technological advances usually coexist with previous innovations, despite various predictions of the end of certain technologies. The telegraph did not take the place of traditional mail. The telephone did not 'kill' the telegraph. Radio coexists with TV. In fact, except for the telegraph, most communication innovations introduced during the last two centuries, including mail, the telephone, radio, TV, and fax, are still in use today. Mail, one of the oldest organised communication methods, is a growing business even today, with large companies like DHL or FedEx.

Radio broadcasting had its renaissance on the internet, with thousands of radio channels broadcast online and podcasts that have gained popularity in the last ten years. Both radio and podcasts are mediums that many people have turned to lately, as they allow people to multitask while listening to the news, their favourite music, or another show.

---

[93] Price, R. B. (1984). Jamming and the Law of International Communications. *Michigan Journal of International Law*, 5(1), 391-403.

[94] United Nations. (1950, December 14). Resolution 424 (V), Freedom of Information: Interference with Radio Signals, UNGA 94. http://www.worldlii.org/int/other/UNGA/1950/95.pdf

Early years of the television

Since its invention in 1926 and the beginning of commercial use six years later by the BBC, television has become a primary news and entertainment medium, and it will keep that position for many years to come. During the 1970s and 1980s, the so-called golden years of television, the idea of 'prime time TV' developed as a new way of addressing a wide audience. For many, it was an integral part of their daily routine. For the first time, we could see and hear world news as it happened. Governments and diplomats started to use television as a quick source of information, and as a powerful tool to convey their messages.

The success and popularity of radio prompted research into the possibility of transmitting moving pictures in a similar way. It is difficult to say who invented television, and exactly when. As had previously happened with the telegraph, with the television, there were a great number of inventors in various countries who worked simultaneously to achieve the same goal during the 1920s.

Two inventors, John Logie Baird from Britain and Charles Francis Jenkins from the United States, built the world's first successful mechanical televisions. By 1929, Baird had managed to produce half-hour shows on the BBC, three times a week. Russian inventor Vladimir Zworykin and American inventor Philo Farnsworth were the pioneers of the electronic TV. Zworkyn invented a workable cathode-ray receiver called the Kinescope, and Farnsworth, with his image dissector camera tube, transmitted the first live human image in 1928.

In 1938, the first electronic television sets became commercially available in the USA. A year later, the opening of the New York World's Fair was broadcast on TV with a speech by President Roosevelt, the first American president to appear on television. By the late 1950s, most European countries had broadcast their own TV programmes and other, less developed nations soon followed.

The contemporaries of the birth of television had different impressions of the new technology. The *New York Times* reviewed a demonstration of television at the 1939 World's Fair:

> The problem with television is that people must sit and keep their eyes glued on a screen; the average American family hasn't time for it.[95]

But J. W. Ridgeway, chairman of the Radio Industry Council from the United Kingdom, said the following:

---

[95] Imagining the Internet https://www.elon.edu/u/imagining/time-capsule/150-years/back-1920-1960/

> It is inevitable that television will become the primary service and sound radio the secondary one.[96]

The next four decades were the golden age of television. In 1963, for the first time, Americans found TV to be a more reliable source of information than print. Prime-time news, reportages, documentaries, and live coverage reached more audiences than any other media had done before. The year 1979 saw the foundation of CNN, the first TV channel devoted to broadcasting news 24 hours a day.

Diplomacy going live: The CNN effect

The possibility of shaping public opinion was most prominent during the first Gulf War in 1991, when the cable news network CNN had reporters and cameras on the ground, reporting the conflict 'live', on television. The CNN team broadcast live from the Rashid Hotel in Baghdad, while no other network could do this. Their coverage was broadcast unedited, with an exciting, dramatic narrative. Real-time news coverage increased transparency, but it complicated sensitive diplomatic relationships between states.

This led some to formulate the theory of the 'CNN effect', a new phenomenon in foreign relations. According to this concept, global television networks play a significant role in determining the actions that diplomats will take, as well as the outcomes of events. The CNN effect shapes public perception, which then affects diplomatic agendas. This theory originated in the 1990s, when CNN covered the interventions in Iraq, Somalia, and Bosnia in real time.

The strong presence of media in a Middle East conflict opened a space for the development of more local broadcasting agencies, such as Al Jazeera, which tries to provide a different perspective on world news.

For diplomats, television has been a source of quickly updated and available information. It also accelerated the speed of diplomatic communication. TV focused world attention on global issues such as terrorism, climate change, and human rights. Leaders sometimes have to address these issues, even when they have not always been among their top priorities. As E. Gilboa writes:

> 'The study of the media's involvement in diplomacy is becoming increasingly important as heads of state and nonstate actors make increasing use of the media as a major instrument for communication and negotiation. [...] In the information age, the inclusion, and sometimes the exclusion, of

---

[96] Elon University. (n.d.). *1920s–1960s: Television.* https://www.elon.edu/u/imagining/time-capsule/150-years/back-1920-1960/

the media from diplomacy will have even more dramatic effects on negotiations. As a greater number of people all over the world watch the same news, leaders and government officials of state and nonstate actors will use the mass media, particularly television, more frequently in both actual negotiations and in the pre-negotiation stages.'[97]

Television is still powerful and important in today's diplomacy. It is important in shaping public opinion, and the general public still relies on traditional television channels as a principal source of information.

The future of TV

Nowadays, television is increasingly streamed over the Internet. It is available on the go, on YouTube or streamed on smartphones and tablets, so viewers can watch news and TV shows whenever and wherever they want. The evolution of television will continue in the coming years. With new technology and new forms of entertainment, television will continue to transform into a personalised experience. Data is being collected based on users' watch history, demographics, and preferences. It helps create smart content, and algorithms determine unique content for each viewer.

In addition, television will probably become immersive. Instead of passively watching, users will be able to participate and interact. Streaming services, like Netflix, are experimenting with interactive content, in which viewers' choices affect the plot. In conclusion, television will probably remain a channel for public diplomacy, even as it changes in the coming years.

Meanwhile...

...in Africa

Although the birth of TV broadcasting in Europe and North America is widely known, the evolution of TV in Africa is less known. The establishment of a Moroccan television station in 1954 marked the beginning of the television age in Africa. In 1959, the Western Nigeria Television Service broadcast the first terrestrial television broadcast signals on the continent. Zimbabwe's first broadcasts came in November 1960, when black-and-white programming started in Harare. Algeria, Kenya, Uganda, and Senegal launched television stations in the late 1950s and early to mid-1960s, but some countries like South Africa and Cameroon didn't have TV stations until the 1970s and 1980s. Nigeria was a front-runner in

---

[97] Gilboa, E. (2001). Diplomacy in the media age: Three models of uses and effects. *Diplomacy and Statecraft* 12(2), 1–28.

introducing news and specific genres of content too, and Nigerian television became the mouthpiece of the government.

Between 1980 and 1985, the Nigerian Television Authority (NTA) started producing Africa's first local soap operas, children's programmes, and comedy series. This marked the birth of the Nollywood film industry, which now produces more than 50 films a week, surpassing Hollywood as the world's second-largest movie industry by number of productions, after India's Bollywood.[98]

In our final chapter, we move into the digital era, to look at the latest technology to have had a transformative effect on diplomacy: the internet. The proliferation of internet communication and social media inaugurated new forms of diplomacy, bringing new methods and platforms of diplomatic interaction.

---

[98] Motsaathebe, G., & Chiumbu, S. H. (2021). *Television in Africa in the Digital Age*. Springer; Zimbojam. (2018, November 21). *Tracing the history of TV in Africa.* https://www.zimbojam.com/world-television-day-discovering-the-critical-role-the-tv/

9. Digital diplomacy: Internet, AI, and social media

'My God, this is the end of diplomacy!'

This was the reported reaction of former British prime minister and foreign secretary Lord Palmerston upon receiving the first telegraph message in the 1850s. More recently, in the 1990s, the US diplomat Zbigniew Brzezinski made the same prediction after using the internet for the first time.[99]

The late 20th and early 21st centuries have seen exponential growth and proliferation of the internet, transforming both mass and personal communication. Many forms of media have been functionally superseded by digital platforms, with much breaking news now being reported first on social media. The spread of mobile internet coverage across the globe has added to this process, with many public figures also engaging with the public directly via social media. The development of the internet, some decades after decolonisation, and accelerating after the end of the Cold War, gave initial hopes that it would be a democratic platform that would permit the free exchange of information. However, the internet has come to be used through the platforms of a small number of corporations, and many states still exert considerable control and surveillance over internet use within (and beyond) their borders, usually in the name of national security.

There have been recent revelations about data collection on social media that have reduced trust in large corporations like Facebook (incurring a rebranding, to Meta) and the platform Twitter, now X, has been acquired by divisive figure Elon Musk. Furthermore, multiple scandals regarding the use and abuse of social media, even by foreign governments, in influencing voting behaviour (most visibly in the 2016 presidential election in the USA), creating ideological echo chambers and spreading the information epidemic of fake news, have further eroded trust in digital communication. Diplomats and the wider practice of diplomacy have had to negotiate these turbulent times.

Diplomacy survived the telephone, radio, television, and many succeeding communications technologies. Although most of these technological innovations influenced diplomacy, they did not challenge the very nature of diplomatic functions. Technology helped diplomats perform their work by offering new tools without altering the

---

[99] Chukwu, O. L. (2018, November 15). Technology is changing diplomacy. *International Policy Digest.* https://intpolicydigest.org/technology-is-changing-diplomacy/

concept of diplomacy. In most cases, it reinforced the importance of diplomacy and defined a more important role for diplomats in society.

Today, diplomats use the internet to find and share information, negotiate, and communicate. Even corridor diplomacy, which was strongly linked to traditional diplomacy, has been replaced by messaging and Twitter (now 'X'). The internet has opened up a two-way communication channel by providing tools for individuals and organisations to influence global policy.

The invention of the internet

The internet started as a government project. In the late 1960s, the US government sponsored the development of the Advanced Research Projects Agency Network (ARPANET), a resilient communications resource designed to survive a nuclear attack. It was a network of computers that would enable government leaders to communicate even in the case of a nuclear attack.[100]

In the mid-1970s, Vinton Cerf, Bob Kahn, and Louis Pousan laid the basis for TCP/IP (Transmission Control Protocol/Internet Protocol). TCP/IP was a way for all computers to communicate with one another even if part of the network was disrupted. The internet continued to evolve, and in 1991, the British computer programmer Tim Berners-Lee invented the World Wide Web: a 'web' of information that anyone on the internet could retrieve. Berners-Lee created the internet that we know today.[101]

Digital diplomacy

Digital diplomacy focuses on the changes in the environment in which diplomacy is practised, the use of internet tools for diplomatic practice, and the new topics on diplomatic agendas, such as privacy, data protection, and cybersecurity.[102]

Diplomats have to deal with a changing landscape of economic and political power, and manage the fast-changing concept of state sovereignty. Future generations of diplomats will have to work in a fundamentally different geopolitical and geo-economic environment.

---

[100] Wikipedia. (n.d.). ARPANET. https://en.wikipedia.org/wiki/ARPANET
[101] Gillies, J., & Cailliau, R. (2000). *How the web was born: The story of the World Wide Web*. Oxford University Press.
[102] Bjola, C., & Holmes, M. (Eds.) (2015). *Digital diplomacy: Theory and practice*. Routledge.

In the United States, diplomats have moved from having diplomatic representatives in, for example, Detroit (the economic hub of the 1950s), to the San Francisco Bay Area (today's economic hub). The newest model of diplomatic representation in the Bay Area is the Office of the Tech Ambassador, introduced by Denmark in 2017. Digital technology will increasingly shape the evolution of the political and economic environment for diplomatic activities.

New topics on diplomatic agendas

The internet brought new topics to diplomatic agendas. These include cybersecurity, data protection, internet governance, and artificial intelligence (AI) governance. In addition to new, digitally driven topics, traditional topics are increasingly influenced by digitalisation. Commerce is becoming e-commerce, health is increasingly digital health, and so on.

On the agenda of the United Nations and its specialised international organisations, there are more and more digital issues. To reflect the emergence of new topics, countries such as Switzerland, the Netherlands, and Australia are developing digital foreign policies. The UN and international organisations are adjusting to this. Many others will follow in the coming years.

Internet as a tool for diplomacy

While conventional forms of diplomacy are still dominant, an increasing number of diplomats use the internet as a new tool for communication, gathering information, and public diplomacy. In the past 20 years, diplomatic tools have changed rapidly: from the introduction of email and the use of websites by diplomatic services and international organisations to the arrival of computers in conference rooms and, most recently, the intensive use of social media such as Facebook and X (formerly Twitter). The introduction of each new e-tool challenged the way things were traditionally done and opened up new opportunities for diplomats and diplomacy.

New tools should help diplomats perform their functions better, as outlined in Article 3 of the Vienna Convention (to represent their countries, negotiate, gather information, and protect the interests of their citizens and companies).[103] Additionally, academic work and debates were steered towards social media and public diplomacy, especially after the Arab Spring, when Facebook and Twitter 'diplomacy' emerged. Fortunately, in recent years, more focus has been put on using digital tools for core diplomatic

---

[103] United Nations. (1961). Vienna Convention on Diplomatic Relations, 18 April 1961. https://legal.un.org/ilc/texts/instruments/english/conventions/9_1_1961.pdf

functions. The COVID-19 pandemic has increased the seriousness of the discussions, in particular with the emergence of online meetings in multilateral diplomacy.

The use of social media in diplomacy

With the expansion of social media, especially Facebook and X/Twitter, diplomats have had to adapt and use social media as additional communication channels.[104] Furthermore, social media proved to be an additional source of information for diplomatic reporting, particularly in environments where diplomatic and media access was limited.

In diplomatic practice, social media can be an important tool for communicating the positions of negotiation parties. The dynamics of the Brexit negotiations, for example, have been shaped by the frequent tweets of chief negotiators and other actors. Social media also enables rapid changes in public opinion. Therefore, diplomats must recognise signals at an early stage and take them very seriously. According to the latest numbers, X/Twitter has close to 400 million users. As users rush to their X/Twitter feeds for the latest news and statements, diplomats have to be prepared. The once-marginal activity of diplomats has become almost a central practice of state diplomacy.[105]

According to Twiplomacy (which provides studies on the use of digital tools by governments and international organisations), in 2018, over 97% of all 193 UN member states had an official presence on Twitter.[106] The most popular platform globally for world leaders is Twitter, followed by Facebook, and then Instagram. Until the 2020 US elections, former US President Donald Trump's personal Twitter account (@realDonaldTrump) was the most followed account globally with 81.8 million followers. Trump was followed by Indian Prime Minister Narendra Modi and Pope Francis. The Pope has over 50 million followers across his nine @Pontifex accounts in multiple languages.

The UN also uses social media platforms and other digital tools to enhance the outreach of its messages. Due to COVID-19, the World Health Organization (WHO) dramatically increased its followers to 21 million on all social media platforms, almost catching up to the UN Children's Fund (UNICEF) and the UN.

---

[104] Clüver Ashbrook, C., & Zalba, A. R. (2021). Social media influence on diplomatic negotiation: Shifting the shape of the table. *Negotiation Journal, 37*(1), 83–96.

[105] Collins, S. D., DeWitt, J. R., & LeFebvre, R. K. (2019). Hashtag diplomacy: twitter as a tool for engaging in public diplomacy and promoting US foreign policy. *Place Branding and Public Diplomacy, 15*(3).

[106] https://www.twiplomacy.com/

DIGITAL DIPLOMACY: INTERNET, AI, AND SOCIAL MEDIA

In 2019, Twiplomacy organised a Twitter poll. They had asked world leaders, governments, and foreign ministries on Twitter how they used the platform and what the benefits of Twitter were for digital diplomacy. Most foreign ministries replied that Twitter is a tool for furthering diplomatic and foreign policy goals, and communicating diplomatic and consular activity. One user wrote that the great benefit of Twitter for diplomacy is that content travels far and fast, directly engaging and impacting people.

Online conferencing and e-participation

COVID-19 has shifted diplomacy online to conferencing platforms such as Zoom. However, online meetings are not as new as one might think. The first remote participation session in multilateral diplomacy was held by the International Telecommunication Union (ITU) in 1963. Internet availability in conference rooms has made remote participation a reality for more inclusive and open international negotiations. In light of COVID-19 social distancing and lockdowns, diplomatic practice has had to adapt. Overall, diplomacy has proven remarkably resilient. Videoconferencing and other

Map of countries with national AI strategies (as on April 2023)

means of digital communication have ensured the continuity of diplomatic practice and negotiations.

However, the absence of informal meeting spaces is regarded as a real loss in terms of relationship building and information gathering. In addition, some protocol requirements do not translate well into online meetings and videoconferencing. In face-to-face meetings, the rank and status of participants are highlighted through arranged seating. However, this type of hierarchy signalling cannot be established as clearly during videoconferencing on standard platforms.

As diplomatic practice has shifted towards videoconferencing, key challenges include solving security issues, adapting to changes in communication and negotiation dynamics, offering translation services, and ensuring a stable internet connection. There are concerns about the danger of exclusion due to bandwidth requirements and security restrictions. This is a particular challenge faced by small and developing countries.

Hybrid (blended) forms of diplomacy that combine in situ and virtual attendance at meetings have emerged as another adaptation. Given the advantages, this form of hybrid diplomacy is here to stay. Diplomatic practice has always been an interplay between continuity and change, and the present moment is a crucial turning point that might determine the future of diplomatic practice.

Metaverse and diplomatic negotiations

The Metaverse, a series of 3D virtual worlds developed by Meta, formerly Facebook, shot into the headlines in 2021. Although the phenomenon is not new, Meta announced their intention to transform the way humans meet and interact digitally through the development of their own virtual world. Tech giants such as Microsoft, Apple, and Google are also developing metaverse applications and tools. Alongside these recent developments, militaries and governments are also reportedly investing heavily in metaverse technologies.

The first glimpse of the emerging metaverse dates back to 2007, when the Maldives established the first virtual embassy in Second Life with the assistance of Diplo. In late 2021, Barbados announced its intention to open a virtual embassy in the metaverse in 2022, while the Israeli embassy in Korea opened a virtual pavilion for embassy events in late 2022.

Meanwhile...

...in Space
Diplomacy isn't just terrestrial anymore. Since the launching of the first satellite into space, diplomats have had to consider the events occurring in the Earth's orbit, on the surface of the Moon, and now even on Mars.

At the beginning of the space age, nations were worried about sending nuclear weapons into orbit, and most of the effort was invested in promoting the peaceful use of outer space. The landmark agreement, the *Outer Space Treaty*, was reached in 1967. It stated that space should only be used for peaceful purposes and that neither space nor celestial bodies could become the sovereign territory of any nation.[107] Also of importance is the so-called *Moon Agreement* of 1979, a multilateral treaty regulating the activities on Earth's only natural satellite, but it was not signed by the USA, the USSR/Russia, or China.[108]

The recent increase in the number of countries and private companies using low Earth orbit (LEO), and our growing dependence on satellites, is increasing the need to negotiate about rights and obligations. There are now more than 3,000 satellites. Experts predict that in ten years, this number will rise to over 100,000. This is not to mention the problem of space debris.

Space is also becoming more commercial and militarised.[109] Even without a new weapons race in orbit, satellite jamming, or destruction, more congestion in space means more potential misunderstandings. The Cold War international law on space is out of date, and new codes of conduct are needed that would oblige states, companies, and individuals to act responsibly in space. The largest legal gaps have to do with the potential commercial use of space. Currently, it is the USA that is shaping space law, and this is why China, Russia, and India remain disengaged. Excluded from the International Space Station for its secretive conduct, China plans to build its own permanent space station.

---

[107] United Nations. (1967). UN Resolution 2222 (XXI), Treaty on Principles Governing the Activities of States in the Exploration and Use of Outer Space, including the Moon and Other Celestial Bodies. https://www.unoosa.org/oosa/en/ourwork/spacelaw/treaties/outerspacetreaty.html

[108] United Nations. (1979). UN Resolution 34/68, Agreement Governing the Activities of States on the Moon and Other Celestial Bodies. https://www.unoosa.org/oosa/en/ourwork/spacelaw/treaties/travaux-preparatoires/moon-agreement.html

[109] Bowen, B. E. (2020). *War in space: Strategy, spacepower, geopolitics.* Edinburgh University Press.

Conclusion

We have looked at the deep, rich, and intertwined histories of diplomacy and technology. As communications technology has changed over time, so too have the methods and practices of diplomacy. Throughout human history, diplomatic practices have consistently adapted to new technological developments, using what they can offer to the diplomat and diplomatic relations between political entities.

While cooperation and conflict resolution were part of human life before writing and communication technologies were developed, their introduction catalysed, as we saw in Chapter 1, other human practices, such as trade, art, gift-giving, and language. Early archaeological examples of writing show the continuity of technological techniques in diplomacy: the diplomatic cable, to this day, remains a cornerstone of diplomatic practice and still takes the form of a written message. While much has changed in the technologies and practices of diplomacy, its dependence on the written word remains as true as ever.

Our knowledge of the ancient world allows us to speak of diplomatic practices in discrete ancient civilisations. In Chapter 2, we discussed ancient Mesopotamian, Egyptian, Assyrian, Persian, Chinese, and Indian diplomatic practices. In each of these civilisations, sophisticated diplomacy was being practised, reminding us that the complexity of the modern world is not without historical precedents.

In Chapter 3, we focused on ancient Greek diplomacy. The intense concentration of linguistically connected city-states in the eastern Mediterranean, and the rich heritage of written accounts about Greek antiquity that we have access to, mean that we can construct detailed reports of the complex diplomatic technologies that were developed and used in this period, such as cryptography and the hydraulic telegraph. As we saw, significant developments were taking place in both China and India during the same period.

Chapter 4 continues with the Greek-speaking world, to look at diplomacy and technology in the Byzantine Empire. This inheritor of the Eastern Roman Empire lasted, astonishingly, for over 1,000 years, an impossible feat without effective diplomacy. The Byzantines developed multiple diplomatic practices that would set the stage for the development of modern diplomacy: a ministry of foreign affairs, the use of soft power, international law, and many more.

In Chapter 5, we moved into the Renaissance and Italy. As it was called in the Western tradition, the 'known world' expanded beyond our imagination in this period. European seafarers established multiple sea routes in this period, such as around the southern tip of Africa into the Indian Ocean and beyond, and over the Atlantic to the Americas. These nautical developments, and subsequent explosions in global trade and colonisation, instigated the beginnings of the geopolitical structure of the modern global world. In the same period, the Italian Peninsula comprised many small city-states, republics, kingdoms, and principalities. This called for intense diplomatic practices, which developed alongside certain world-altering technological advancements, such as the invention of the printing press, making mass production of the written word possible for the first time. The Renaissance period set essential precedents for diplomacy and technology in future years.

Chapter 6 saw us move into what we know as the modern world. The Peace of Westphalia in Europe laid the groundwork for the emergence of modern sovereignty, a key principle of international relations and statecraft to this day. Alongside the expansion of multiple European empires, such as those of the British and French, the invention of the telegraph and the laying of deep-sea telegraph cables made for the beginnings of a globally interconnected world, with diplomats and state officials able to take advantage of fast communication, if not all at the same time, as some developed and consolidated this technology faster than others.

In Chapter 7, we saw the development of the telephone in the late 19th century. While the telegraph enabled the electric transmission of coded messages, the phone allowed instantaneous conversations. The telephone would go on to play a crucial role in many significant international and diplomatic crises, such as the Six-Day War of 1967, and the inadequacy of communication between the United States and the Soviet Union during the Cuban Missile Crisis led to the development of a direct telephone connection between the two Cold War giants.

In Chapter 8, we went wireless for the first time with the invention of radio and TV broadcasting, accelerating the phenomenon of public diplomacy. By the mid-20th century, the number of privately owned wireless radios was exploding, as was ownership of television sets, bringing sound and vision into the home. The radio had been used for public diplomacy since its introduction into consumer markets, and the blocking of radio signals was a feature of Cold War border regions in Eastern Europe as each side attempted to limit the propaganda efforts of their adversary. This issue would reach the UN by 1950.

Finally, as consumer electronics became more advanced and the internet became an accessible phenomenon to more and more people, we discussed diplomacy in the internet and social media age in Chapter 9. Public diplomacy has moved to the forefront, with diplomats and embassies overseeing social media accounts intended to disseminate directly to the broadest possible audience. Alongside public diplomacy, the internet age has also introduced new means of communication between diplomats

CONCLUSION

and embassies and new diplomatic issues, such as internet governance and regulation. While traditional diplomatic practices remain, the internet has inaugurated a new age of instant communication and the development of a vast audience for the practice of public diplomacy.

AI ahead of us

AI will profoundly impact diplomacy in three main areas: the impact of AI on geopolitics, the question of AI governance on the diplomatic agenda, and the use of AI tools in the practice of diplomacy.

In geopolitics, AI will impact the power distribution and position of countries in future military and economic relations. It is the reason why many countries are positioning AI as a strategic asset with major investments in academic research and economic development. As of April 2023, 58 countries have national AI strategies, while 25 are developing them.

While AI is mostly being developed by a few wealthy and powerful organisations and governments, there are also signs of its development in other parts of the world. In Africa, for example, we see AI development taking place in Egypt, Nigeria, Kenya, and South Africa, and Mauritius. Ethiopia, Ghana, Rwanda, and Uganda are all working to encourage AI research and development. The African Union, too, is working on a pan-African AI strategy,

AI governance is rising in relevance on the diplomatic agenda. Governments, tech companies, and international organisations are calling for international rules on AI development. They range from overall AI governance to regulating the specific impact of AI on, among other things, privacy, e-commerce, cybersecurity, disarmament, and intellectual property. The main negotiations on AI governance will happen in the context of the UN Global Digital Compact, the EU's AI Act, and the OECD Recommendations on AI.

Lastly, AI will profoundly impact diplomatic practice. For example, it will automate a considerable part of diplomatic reporting, which is one of the main activities of diplomats. AI will also help draft resolutions and diplomatic documents, and diplomatic services will use AI for planning and scenario development.

While the metaverse, AI, and other emerging technologies will bring many new developments, their impact on diplomacy will likely emerge through the continuity of the core functions of diplomacy and changes in the way diplomacy is conducted.

Bibliography

Further reading links:

'Acheulean', in *Wikipedia*, n.d., https://en.wikipedia.org/wiki/oldowan.

"Antonio Meucci" Engineering and Technology History Wiki, n.d. https://ethw.org/Antonio_Meucci

ARPANET, Wikipedia, (n.d.) https://en.wikipedia.org/wiki/ARPANET

'Aztec Warfare', World History Encyclopedia, n.d., https://www.worldhistory.org/Aztec_Warfare/.

"Gartner hype cycle," Wikipedia: The Free Encyclopedia. San Francisco: Wikimedia Foundation, https://en.wikipedia.org/wiki/Gartner_hype_cycle

"Imagining the Internet: A History and Forecast, 1870s – 1940s: Telephone", Elon University, n.d. https://www.elon.edu/u/imagining/time-capsule/150-years/back-1870-1940/

Lily Filson, 'Special Topics Lecture 4: Byzantine Technology," https://filsonarthistory.wordpress.com/2019/01/16/special-topics-lecture-4-byzantine-technology/

'Oldowan', in *Wikipedia*, n.d., https://en.wikipedia.org/wiki/Acheulean.

"Tracing the history of TV in Africa", *ZimboJam*, November 21, 2018. https://www.zimbojam.com/world-television-day-discovering-the-critical-role-the-tv/

"The Washington-Moscow Hotline", *Electrospaces*, October 28, 2012. https://www.electrospaces.net/2012/10/the-washington-moscow-hot-line.html

"1920s – 1960s: Television", Elon University, n.d. https://www.elon.edu/u/imagining/time-capsule/150-years/back-1920-1960/

https://www.twiplomacy.com/

https://www.youtube.com/watch?v=Az28Hsne-nI

Images

A relief of the Greek hydraulic telegraph of Aeneas, depicting one half of a telegraph system. Public Domain, Wikimedia Commons. https://commons.wikimedia.org/wiki/File:Greek_Hydraulic_Telegraph_of_Aeneas_relief.jpg

Antonio Meucci, (1808 - 1889) inventor of the telephone, Public Domain, https://commons.wikimedia.org/wiki/File:Antonio_Meucci.jpg

Chappe's telegraph, Illustration appeared in 'Les merveilles de la science', Louis Figuier, 1868, Public Domain, https://commons.wikimedia.org/wiki/File:T%C3%A9l%C3%A9graphe_Chappe_1.jpg

Cristofano dell'Altissimo. *Portrait of Pope Alexander VI*. Vasari Corridor. Public Domain. https://commons.wikimedia.org/wiki/File:Pope_Alexander_Vi.jpg.

Delegates of the Congress of Vienna in a contemporary engraving by Jean Godefroy after the painting by Jean-Baptiste Isabey, CC BY-SA 3.0. https://commons.wikimedia.org/wiki/File:Congress_of_Vienna.PNG

Descouens, Didier. *An Acheulean Handaxe, Haute-Garonne France*. Lower Paleolithic - Acheulan. Chert, Muséum de Toulouse. CC BY-SA 4.0 https://commons.wikimedia.org/wiki/File:Biface_Cintegabelle_MHNT_PRE_2009.0.201.1_V2.jpg.

Enok, Map of Italy 1494. CC BY-SA 3.0 https://commons.wikimedia.org/wiki/File:Italy_1494_de.svg#/media/File:Italy_1494.svg

Gobierno de Cantabria, Altamira Cave Painting, CC BY 3.0. https://commons.wikimedia.org/wiki/File:Altamira-2.jpg

Guglielmo Marconi, portrait, 1908, Public Domain, https://commons.wikimedia.org/wiki/File:Guglielmo_Marconi.jpg

Hans Holbein the Younger, *The Ambassadors*. 1533. Public Domain. https://en.wikipedia.org/w/index.php?title=File:Hans_Holbein_the_Younger_-_The_Ambassadors_-_Google_Art_Project.jpg&oldid=679553176.

J. Siebold. Waldseemüller map, 1507. Public Domain https://commons.wikimedia.org/wiki/File:Waldseemuller_map_in_color.jpg

Jastrow. So-called "Logios Hermes" (Hermes,Orator). Marble, Roman copy from the late 1st century CE-early 2nd century CE after a Greek original of the 5th century BC, by Phidias. Public Domain, Wikimedia Commons https://commons.wikimedia.org/wiki/File:Hermes_Logios_Altemps_33.jpg

Joeykentin. *Bone Tool and Possible Mathematical Device That Dates to the Upper Paleolithic Era Discovered in Ishango*. Own work. CC BY-SA 4.0 https://commons.wikimedia.org/wiki/File:Ishango_bone.jpg#filelinks

Library of Ashurbanipal Mesopotamia 1500-539 BC Gallery, British Museum, Public Domain https://commons.wikimedia.org/wiki/File:Library_of_Ashurbanipal.jpg

Louvre Museum, Stele of Hammurabi, circa 1793-1751 BC, CC BY 3,0. https://commons.wikimedia.org/wiki/File:F0182_Louvre_Code_Hammourabi_Bas-relief_Sb8_rwk.jpg

Marsyas. Carte du monde égéen en 431 av. J.-C., à la veille de la Guerre du Péloponnèse. CC BY-SA 3.0. https://commons.wikimedia.org/wiki/Atlas_of_Greece#/media/File:Map_Peloponnesian_War_431_BC-fr.svg

Museum of the Ancient Orient, Table of Treaty of Kadesh from Boğazköy, Turkey. CC BY 3.0. https://commons.wikimedia.org/wiki/File:Table_of_Treaty_of_Kadesh_from_Bo%C4%9Fazk%C3%B6y.jpg

BIBLIOGRAPHY

Nafsadh, Map of the Fertile Crescent, CC BY-SA 4.0, https://commons.wikimedia.org/wiki/File:Map_of_fertile_crescent.svg

Osama Shukir Muhammed Amin, Five Amarna letters on display (G57/dc8) at the British Museum, London, CC BY-SA 4.0 https://commons.wikimedia.org/wiki/File:Five_Amarna_letters_on_display_at_the_British_Museum,_LondonA.jpg

Printmaker, Larmessin, Nicolas de, (1632-1694). Johannes Gutenberg. Wikimedia Commons, Public Domain. https://commons.wikimedia.org/wiki/File:Johannes_Gutenberg.jpg

Pyrseia code, Public Domain, Wikimedia Commons. https://commons.wikimedia.org/wiki/File:Polybius_square.png

Shakko. Cast of a Roman copy of the original early 4th Century bust of Thucydides. CC BY-SA 3.0, Wikimedia Commons. https://commons.wikimedia.org/wiki/File:Thucydides_pushkin02.jpg

Sting, Bust of the Greek orator Demosthenes. Marble, Roman artwork, inspired by a bronze statue by Polyeuctos (ca. 280 BC). Found in Italy. CC BY-SA 2.5, Wikimedia Commons. https://commons.wikimedia.org/wiki/File:Demosthenes_orator_Louvre.jpg

Studio of Michiel Jansz van Mierevelt. *Sir Henry Wotton (1568-1639)*. 1620. Sotheby's. Public Domain. https://commons.wikimedia.org/wiki/File:Sir_Henry_Wotton_(1568-1639),_Studio_of_Michiel_Jansz_van_Mierevelt.jpg.

Thoth, Public Domain, https://commons.wikimedia.org/wiki/File:The_Sacred_Books_and_Early_Literature_of_the_East,_vol._2,_pg._416-417,_Thoth.jpg

Transatlantic telephone cable routes under study in early 1956. American Telephone and Telegraph Company. From J. S. Jac, W. H. Leech, & H. A. Lewis, "Route Selection and Cable Laying for the Transatlantic Cable System," Bell System Technical Journal, I (1957) No. 1, 296. Flickr Commons, 'No known copyright restrictions' license, https://commons.wikimedia.org/wiki/File:The_Bell_System_technical_journal_(1922)_(14753547484).jpg

Winston Churchill at a BBC microphone about to broadcast to the nation on the afternoon of VE Day, 8 May 1945. Source: Imperial War Museum, Public Domain, https://en.m.wikipedia.org/wiki/File:Winston_Churchill_at_a_BBC_microphone_about_to_broadcast_to_the_nation_on_the_afternoon_of_VE_Day,_8_May_1945._H41843.jpg

Zimmermann Telegram as Received by the German Ambassador to Mexico , Public Domain, https://commons.wikimedia.org/wiki/File:Zimmermann_Telegram_as_Received_by_the_German_Ambassador_to_Mexico_-_NARA_-_302025.jpg

Treaties

The League of Nations, Treaty 4319, The International Convention Concerning the Use of Broadcasting in the Cause of Peace, LNTSer 80; 186 LNTS 301, (September 23, 1936), http://www.worldlii.org/int/other/treaties/LNTSer/1938/80.html

United Nations, Resolution 424 (V), Freedom of Information: Interference with Radio Signals, UNGA 94, December 14 1950. http://www.worldlii.org/int/other/UNGA/1950/95.pdf

United Nations, Vienna Convention on Diplomatic Relations, April 18 1961. https://legal.un.org/ilc/texts/instruments/english/conventions/9_1_1961.pdf

UN Treaty 6791, International Coffee Agreement, July 1, 1963. https://treaties.un.org/doc/Publication/UNTS/Volume%20469/v469.pdf

UN Resolution 2222 (XXI), Treaty on Principles Governing the Activities of States in the Exploration and Use of Outer Space, including the Moon and Other Celestial Bodies, 1967. https://www.unoosa.org/oosa/en/ourwork/spacelaw/treaties/outerspacetreaty.html

UN Resolution 34/68, Agreement Governing the Activities of States on the Moon and Other Celestial Bodies, 1979. https://www.unoosa.org/oosa/en/ourwork/spacelaw/treaties/travaux-preparatoires/moon-agreement.html

Citations

Aeschylus. *Agamemnon*, edited by Eduard Fraenkel. Oxford University Press, 1962.

Antl-Weiser, Walpurga. "The Time of the Willendorf Figurines and New Results of Palaeolithic Research in Lower Austria." *Anthropologie (1962-)* 47, no. 1/2 (2009): 131–41.

Ashbrook, Cathryn Clüver and Alvaro Renedo Zalba. "Social Media Influence on Diplomatic Negotiation: Shifting the Shape of the Table," *Negotiation Journal* 37, no. 1 (Winter 2021): 83-96.

Austin, N. J. E. and N. B. Rankov, *Exploratio: Military & Political Intelligence in the Roman World from the Second Punic War to the Battle of Adrianople.* London: Routledge, 1998.

Bates, Darrell. *The Fashoda Incident of 1898: Encounter on the Nile.* Oxford University Press, 1984.

BBC, "Winston Churchill's first wartime broadcast", 1 October 1939. https://www.bbc.com/historyofthebbc/anniversaries/october/winston-churchills-first-wartime-broadcast

Bell, Lanny. "Conflict and Reconciliation in the Ancient Middle East: The Clash of Egyptian and Hittite Chariots in Syria, and the World's First Peace Treaty between "Superpowers"." in *War and Peace in the Ancient World*, edited by Kurt A. Raaflaub (John Wiley & Sons, 2008), 98–120.

BIBLIOGRAPHY

Bergmeier, H. J. P. and Rainer E. Lotz. *Hitler's Airwaves: The Inside Story of Nazi Radio Broadcasting and Propaganda Swing, Volume 1.* Yale University Press, 1997.

Berridge, Geoff. *The Diplomacy of Ancient Greece - A Short Introduction.* DiploFoundation, 2018: https://issuu.com/diplo/docs/the_diplomacy_of_ancient_greece.

Bertman, Stephen. *Handbook to Life in Ancient Mesopotamia.* Oxford University Press, 2005.

Biser, Margaret. "The Fireside Chats: Roosevelt's Radio Talks" *The White House Historical Association*, n.d. https://www.whitehousehistory.org/the-fireside-chats-roosevelts-radio-talks

Bjola, Corneliu and Marcus Holmes, eds. *Digital Diplomacy: Theory and Practice,* (Routledge, 2015)

Boissoneault, Lorraine. "Colored Pigments and Complex Tools Suggest Humans Were Trading 100,000 Years Earlier Than Previously Believed." Smithsonian Magazine, 15 March 2018, https://www.smithsonianmag.co/ture/colored-pigments-and-complex-tools-suggest-human-trade-100000-years-earlier-previously-believed-180968499/.m/science-na

Bowen, Bleddyn E. *War in Space: Strategy, Spacepower, Geopolitics.* Edinburgh University Press, 2020.

Brett, Gerard. "The Automata in the Byzantine "Throne of Solomon."" *Speculum* 29, no. 3 (1954).

Brumm, Adam et al. "Oldest Cave Art Found in Sulawesi." *Science Advances* 7, no. 3 (13 January 2021). https://doi.org/10.1126/sciadv.abd4648.

Bryant, Peter. *Red Alert.* Ace Books, 1958.

Campbell, W. Joseph. *Yellow Journalism: Puncturing the Myths, Defining the Legacies.* Praeger, 2001.

Childress, Diana. *Johannes Gutenberg and the Printing Press.* Minneapolis: Twenty-First Century Books, 2008.

Chukwu, Obinnaya Lucian. "Technology is Changing Diplomacy." *International Policy Digest*, November 15, 2018. https://intpolicydigest.org/technology-is-changing-diplomacy/

Cohen, Raymond. "Diplomacy through the Ages", in *Diplomacy in a Globalizing World: Theories and Practices*, edited by Pauline Kerr and Geoffrey Wisema. Oxford University Press, 2018, 15–20.

Cohen, Raymond and Raymond Westbrook. *Amarna Diplomacy: The Beginnings of International Relations.* JHU Press, 2002.

Collins, Stephen D., Jeff R. DeWitt, Rebecca K. LeFebvre. "Hashtag diplomacy: Twitter as a tool for engaging in public diplomacy and promoting US foreign policy." *Place Branding and Public Diplomacy* 15, no. 3 (2019).

Constantine Porphyrogennetos. *The Book of Ceremonies*. Translated by Ann Moffatt and Maxeme Tall, Brill, 2017.

Dawson, Keith. "Browser Wars of the Wireless Tepgraphy Age." *TBTF*, January 3, 1999 https://tbtf.com/resource/telegraph-browser-wars.html

Dell'Amore, Christine. "Prehistoric Americans Traded Chocolate for Turquoise?" National Geographic News, 29 March 2011, https://www.nationalgeographic.com/history/article/110329-chocolate-turquoise-trade-prehistoric-peoples-archaeology.

Dillon, Wilton S. *Gifts and Nations: The Obligation to Give, Receive and Repay*. Routledge, 2017.

Diplo. (2019). Mapping the challenges and opportunities of artificial intelligence for the conduct of diplomacy. https://www.diplomacy.edu/wp-content/uploads/2019/02/AI-diplo-report.pdf

Diplo. (2022). Stronger digital voices from Africa: Building African digital foreign policy and diplomacy. https://www.diplomacy.edu/wp-content/uploads/2022/11/Stronger-digital-voices-from-Africa.pdf

'Diplomacy - History of Diplomacy | Britannica', n.d., https://www.britannica.com/topic/diplomacy/History-of-diplomacy."

Dixson, Alan F. and Barnaby J. Dixson. "Venus Figurines of the European Paleolithic: Symbols of Fertility or Attractiveness?" *Journal of Anthropology* 2011 (2011): 1–11.

Ferguson, Niall. *The House of Rothschild: The World's Banker 1849-1998*. Penguin, 2000.

Gilboa, Eytan. "Diplomacy in the media age: Three models of uses and effects." *Diplomacy and Statecraft* 12, no. 2 (2001): 1-28.

Gillies, James and Robert Cailliau. *How the Web was Born: The Story of the World Wide Web*. Oxford University Press, 2000.

J. A. J. Gowlett. "The Discovery of Fire by Humans: A Long and Convoluted Process." *Philosophical Transactions of the Royal Society B: Biological Sciences* 371, no. 1696 (5 June 2016), https://doi.org/10.1098/rstb.2015.0164..

Greer, Robert. "Hedonism, absinthe and Parisian decadence: La Belle Époque." *ArtUK*, April 18, 2017. https://artuk.org/discover/stories/hedonism-absinthe-and-parisian-decadence-la-belle-poque

Grotius, Hugo. *On the Law of War and Peace*. Jazzybee Verlag, 2018.

Grotius, Hugo. *The Free Sea*, ed. Richard Hakluyt and William Welwood. Liberty Fund, 2004.

Hobbes, Thomas. *Leviathan*. Oxford University Press, 1996.

Hung, Hsiao-Chun. et al. "Ancient Jades Map 3,000 Years of Prehistoric Exchange in Southeast Asia" *Proceedings of the National Academy of Sciences* 104, no. 50 (11 December 2007): 19745–50, https://doi.org/10.1073/pnas.0707304104.

ITU, 'The Earliest International Telegraph Agreements', *ITU Portal* http://handle.itu.int/11.1004/020.2000/s.139

BIBLIOGRAPHY

Jaspers, Karl. *The Origin and Goal of History*. Yale University Press, 1953.

Johnston, Douglas M. *The Historical Foundations of World Order: The Tower and the Arena*. Martinus Nijhoff Publishers, 2008.

Gheverghese Joseph, George. *The Crest of the Peacock: Non-European Roots of Mathematics*. Princeton University Press, 2011.

Guicciardini, Francesco. *Guicciardini's Ricordi: The Counsels and Reflections of Francesco Guicciardini*, edited by. G. R. Berridge. Leicester, 2000.

Kahn, David. 'The History of Steganography', in *Information Hiding: First International Workshop, Cambridge, U.K., May 30 - June 1, 1996. Proceedings*, ed. Ross Anderson (Springer, 1996), 1–5.

Kahneman, Daniel. *Thinking, Fast and Slow*. Farrar, Straus and Giroux, 2011.

Kauṭalya. *The Arthashastra*. Penguin Books India, 1992.

Kern, Stephen. The Culture of Time and Space, 1880-1918. Harvard University Press, 1983.

Kurbalija, Jovan. 'What Can Diplomacy Learn from Primates? [Podcast Interview with Prof. Frans de Waal] - Diplo', 11 February 2021, https://www.diplomacy.edu/blog/interview-prof-frans-de-waal-what-can-diplomacy-learn-primates/.

Krishnan, Revathi. "J.C. Bose – 'Father of Radio Science' who was forgotten by West due to his aversion to patents," *ThePrint*, November 30, 2020. https://theprint.in/features/j-c-bose-father-of-radio-science-who-was-forgotten-by-west-due-to-his-aversion-to-patents/552556/

Leroi-Gourhan, Arlette. "The Archaeology of Lascaux Cave." *Scientific American* 246, no. 6 (1982): 104–13.

Marsh, Alison. Inside the Third Reich's Radio, *IEEE Spectrum*, March 30, 2021 https://spectrum.ieee.org/inside-the-third-reichs-radio

McClellan, James E. and Harold Dorn. *Science and Technology in World History*. The Johns Hopkins University Press, 2006.

Van De Mieroop, Marc. *King Hammurabi of Babylon: A Biography*. John Wiley & Sons, 2008.

Mosco, Vincent. *The Digital Sublime: Myth, Power, and Cyberspace*. MIT Press, 2005.

Motsaathebe, Gilbert and Sarah H. Chiumbu. *Television in Africa in the Digital Age*. Springer, 2021.

Nickles, David Paull. *Under the Wire: How the Telegraph Changed Diplomacy*. Harvard University Press, 2003.

Nicolson, Harold. *Diplomacy*. Oxford University Press, 1950.

Nicolson, Harold. *The Evolution of Diplomacy*. Collier Books, 1962.

Plato, *Phaedrus* (Penguin Publishing Group, 2005).

Pooley, Jefferson and Michael J. Socolow, "The Myth of the *War of the Worlds* Panic, *Slate*", October 28, 2013. https://slate.com/culture/2013/10/orson-welles-war-of-the-worlds-panic-myth-the-infamous-radio-broadcast-did-not-cause-a-nation-wide-hysteria.html

Price, Rochelle B. "Jamming and the Law of International Communications." *Michigan Journal of International Law* 5, no. 1 (1984), 391-403.

Ramm, Agatha, ed. *The Gladstone-Granville Correspondence*. Cambridge University Press, 1998.

Ramos, Pedro A. Saura. *Cave of Altamira*. Harry N. Abrams, 1999.

Roosevelt, Franklin Delano. *The Fireside Chats of Franklin Delano Roosevelt*. Project Gutenberg, 2003.

Sasson, Jack M. *From the Mari Archives: An Anthology of Old Babylonian Letters*. Pennsylvania State University Press, 2017.

Schwartz, A. Brad. *Broadcast Hysteria: Orson Welles's War of the Worlds and the Art of Fake News*. Hill and Wang, 2015.

Schweikard, David P. and Hans Bernhard Schmid, 'Collective Intentionality', in *The Stanford Encyclopedia of Philosophy*, edited by Edward N. Zalta, Fall 2021 (Metaphysics Research Lab, Stanford University, 2021), https://plato.stanford.edu/archives/fall2021/entries/collective-intentionality/

Sabine, Robert. *The Electric Telegraph*. Virtue, 1867.

Séfériadès, Michel Louis. "Spondylus and Long-Distance Trade in Prehistoric Europe." in *The Lost World of Old Europe: The Danube Valley, 5000-3500 BC*, edited by David W. Anthony and Jennifer Chi. Princeton University Press, 2010, 179–89.

Seland, Eivind Heldaas. "Archaeology of Trade in the Western Indian Ocean, 300 BC–AD 700." *Journal of Archaeological Research* 22, no. 4 (1 December 2014): 367–402, https://doi.org/10.1007/s10814-014-9075-7.

Seri, Andrea. "Adaptation of Cuneiform to Write Akkadian" in *Visible Language: Inventions of Writing in the Ancient Middle East and Beyond*, edited by Woods Christopher. Oriental Institute of the University of Chicago, 2010, 85–93.

Sherrow, Victoria. *Guglielmo Marconi: Inventor of Radio and Wireless Communication*. Enslow Publishers, 2004.

Sikora, Martin et al. "Ancient Genomes Show Social and Reproductive Behavior of Early Upper Paleolithic Foragers." *Science* 358, no. 6363 (3 November 2017): 659–62, https://doi.org/10.1126/science.aao1807.

Sowerby, Tracy. "The Role of the Ambassador and the use of Ciphers." *State Papers Online* 1509–1714

Standage, Tom. *The Victorian Internet: The Remarkable Story of the Telegraph and the Nineteenth Century's On-Line Pioneers*. Berkley Books, 1999.

BIBLIOGRAPHY

Taylor, L. (2023, May 2). Amnesty International criticised for using AI-generated images. *The Guardian.* https://www.theguardian.com/world/2023/may/02/amnesty-international-ai-generated-images-criticism

Ulm Museum. *The Return of the Lion Man: History - Myth - Magic.* Thorbecke, 2013.

UNCTAD. (2018, October 15). Small economies welcome AI-enabled trade tool, but worries remain. https://unctad.org/news/small-economies-welcome-ai-enabled-trade-tool-worries-remain

Wachter, Karl. "Uncovering Prehistoric Danube Culture | ICPDR - International Commission for the Protection of the Danube River." *Danube Watch* 2 (2013), https://www.icpdr.org/main/publications/uncovering-prehistoric-danube-culture.

Weber, Ralph E. "Seward's other folly: America's first encrypted cable." *Cryptologic Quarterly* 12, no. 2 (1993): 81-103.

Whitehead, Alfred North. *Process and Reality.* The Free Press, 1978.

Woods, Christopher. "The Earliest Mesopotamian Writing." in *Visible Language: Inventions of Writing in the Ancient Middle East and Beyond*, edited by Christopher Woods. Oriental Institute of the University of Chicago, 2010, 33–50.

Wurz, Sarah. "The Transition to Modern Behavior." *Nature Education Knowledge* 3, no. 10 (2012).

Xenophon. *Cyropaedia.* translated by Walter Miller. William Heinemann, 1961.

About the author

Dr Jovan Kurbalija is the Executive Director of DiploFoundation and Head of the Geneva Internet Platform (GIP). He was a member of the UN Working Group on Internet Governance (2004/2005), special advisor to the Chairman of the UN Internet Governance Forum (2006–2010), and a member of the High-Level Multistakeholder Committee for NETmundial (2013/2014). In 2018/2019, he served as Co-Executive Director of the Secretariat of the United Nations High-Level Panel on Digital Cooperation.

A former diplomat, Jovan has a professional and academic background in international law, diplomacy, and information technology. He has been a pioneer in the field of cyber diplomacy since 1992 when he established the Unit for Information Technology and Diplomacy at the Mediterranean Academy of Diplomatic Studies in Malta, and later, DiploFoundation.

Since 1997, Jovan's research and articles on cyber diplomacy have shaped research and policy discussions on the impact of the internet on diplomacy and international relations. He lectures on e-diplomacy and internet governance in academic and training institutions in many countries, including Austria (Diplomatic Academy of Vienna), Belgium (College of Europe), Switzerland (University of St Gallen), Malta (University of Malta), and the United States (University of Southern California).

Jovan has published and edited numerous books, articles, and chapters, including *The Internet Guide for Diplomats, Knowledge and Diplomacy, The Influence of IT on Diplomatic Practice, Information Technology and the Diplomatic Services of Developing Countries, Modern Diplomacy,* and *Language and Diplomacy*. With Stefano Baldi and Eduardo Gelbstein, he co-authored the *Information Society Library*, a set of eight booklets covering a wide range of internet-related developments. His book *An Introduction to Internet Governance* provides a comprehensive overview of the main issues and actors in the field through a practical framework for analysis, discussion, and resolution of significant issues. With its seven editions, the book has been translated into 11 languages and it was selected as the Book of the Month for December 2017 by the United Nations Library in Geneva.